Scalable Data Architecture with Java

Build efficient enterprise-grade data architecting solutions using Java

Sinchan Banerjee

BIRMINGHAM—MUMBAI

Scalable Data Architecture with Java

Publishing Product Manager: Ali Abidi
Senior Editor: Nazia Shaikh
Content Development Editor: Manikandan Kurup
Technical Editor: Sweety Pagaria
Copy Editor: Safis Editing
Project Coordinator: Farheen Fathima
Proofreader: Safis Editing
Indexer: Rekha Nair
Production Designer: Shyam Sundar Korumilli
Marketing Coordinator: Abeer Dawe

First published: October 2022

Production reference: 1220922

Published by Packt Publishing Ltd.
Livery Place
35 Livery Street
Birmingham
B3 2PB, UK.

ISBN 978-1-80107-308-0

www.packt.com

Dedicated to Maa and Baba, to whom I am eternally thankful for the way they raised me to become the person I am today

Contributors

About the author

Sinchan Banerjee is a principal data architect at UST Inc. He works for their client Anthem to architect, build, and deliver scalable, robust data engineering solutions to solve their business problems. Prior to his journey with UST, he worked for various Fortune 500 organizations, such as Amex, Optum, Impetus, and HP, designing, architecting, and building robust data engineering solutions for very high volumes of data. He is the lead author of a patent on storage capacity forecasting and is the co-author of multiple international publications. He is also a certified AWS Professional and a certified Java programmer. He has also been a recipient of multiple awards and accolades for exceptional technical contribution, leadership, and innovation.

I would like to thank Packt publication for giving me the opportunity to write this book and share my knowledge. I am also grateful to the editorial and technical reviewer team for their valuable inputs and reviews which made the book a better read. Finally, I would like to thank my wife, who partnered this journey of book writing along with me and constantly inspired, supported, and encouraged me to write to the best of my abilities.

About the reviewers

Sourin Sarkar started his journey as a programmer almost 30 years ago. Currently, he is an architect in a top memory technology company. He works on various security solutions in the embedded security space. He has worked in various technology areas during his career and has been architecting technology solutions for the past 15 years while working with technology giants in the security, memory, storage, and data center domains. He is enthusiastic about security technology, various macro to nano embedded technology, automotive solutions, autonomous solutions, memory technology, robotics, green technology sectors, and various other technologies. He is active in the innovation space and has many issued and filed patents to his name.

I would like to thank my parents, teachers, friends, and highly respected mentors in my career for what I am today. Without them, it would not have been possible. I would like to thank Packt Publishing for giving me this opportunity to review a wonderful book and wish all the best to the author, while looking forward to working with Packt Publishing in the future.

Khushboo K is a big data leader with over a decade of IT experience. She has led and delivered several data engineering solutions for various clients in the US, UK, India, and New Zealand. After a stint with multiple multinational corporations, she started her own venture where she provides data engineering consultancy and training services.

Table of Contents

Section 2 – Building Data Processing Pipelines

4

ETL Data Load – A Batch-Based Solution to Ingesting Data in a Data Warehouse   81

5

Architecting a Batch Processing Pipeline   119

6

Architecting a Real-Time Processing Pipeline 149

7

Core Architectural Design Patterns 181

8

Section 3 – Enabling Data as a Service

9

10

Section 4 – Choosing Suitable Data Architecture

11

12

Preface

When I started writing this book, I looked back at my experience in architecting and developing data engineering solutions, delivering and running those solutions effectively in production, and helping many companies to build and manage scalable and robust data pipelines and asked myself – *What are the most useful things that I can share to help an aspiring or beginner data architect, a data engineer, or a Java developer to become an expert data architect?* This book reflects the work I do on a daily basis, to design, develop, and maintain scalable, robust, and cost-effective solutions for different data-engineering problems.

Java architectural patterns and tools enable architects to develop reliable, scalable, and secure data engineering solutions to collect, manipulate, manage, and publish data. There are many books and online materials that discuss data architectures in general. There are other sets of books and online materials that focus on and dive deep into the technology stack. While such materials provide architects with essential knowledge, they often lack details on how an architect should approach a data engineering problem practically and create the best-suited architecture by using logical inference. In this book, I have tried to formalize a few techniques by which a data architect can approach a problem to create effective solutions.

In this book, I will take you on a journey in which you learn the basics of data engineering and how to use the basics to analyze and propose solutions for a data engineering problem. I also discuss how a beginner architect can choose the correct technology stack to implement a solution. I also touch upon data security and governance for those solutions.

One of the challenges that architects face is there is always more than one way to do things. We also discuss how to measure different architectural alternatives and how you can correctly choose the best-suited alternative using data-driven techniques.

Who this book is for

Scalable Data Architecture is written for Java developers, data engineers, and aspiring data architects who have at least some working knowledge of either backend systems or data engineering solutions. This book assumes that you have at least some working knowledge of Java and know the basic concepts of Java. This book will help you grow into a successful Java-based data architect.

Data architects and associate architects will find this book helpful to hone their skills and excel at their work. Non-Java backend developers or data engineers can also use the concepts of this book. However, it might be difficult for them to follow the code and implementation of the solutions.

What this book covers

Chapter 1, Basics of Modern Data Architecture, is a short introduction to data engineering, basic concepts of data engineering, and the role a Java data architect plays in data engineering.

Chapter 2, Data Storage and Databases, is a brief discussion about various data types, storage formats, data formats, and databases. It also discusses when to use them.

Chapter 3, Identifying the Right Data Platform, provides an overview of various platforms to deploy data pipelines and how to choose the correct platform.

Chapter 4, ETL Data Load – A Batch-Based Solution to Ingest Data in a Data Warehouse, discusses how to approach, analyze, and architect an effective solution for a batch-based data ingestion problem using Spring Batch and Java.

Chapter 5, Architecting a Batch Processing Pipeline, discusses how to architect and implement a data analysis pipeline in AWS using S3, Apache Spark (Java), AWS **Elastic MapReduce** (**EMR**), and AWS Athena for a big data use case.

Chapter 6, Architecting a Real-Time Processing Pipeline, provides a step-by-step guide to building a real-time streaming solution to predict the risk category of a loan application using Java, Kafka, and related technologies.

Chapter 7, Core Architectural Design Patterns, discusses various common architectural patterns used to solve data engineering problems and when to use them.

Chapter 8, Enabling Data security and Governance, introduces data governance and discusses how to apply it using a practical use case. It also briefly touches upon the topic of data security.

Chapter 9, Exposing MongoDB Data as a Service, provides a step-by-step guide on how to build Data as a Service to expose MongoDB data using a REST API.

Chapter 10, Federated and Scalable DaaS with GraphQL, discusses what GraphQL is, various GraphQL patterns, and how to publish data using GraphQL.

Chapter 11, Measuring Performance and Benchmarking Your Applications, provides an overview of performance engineering, how to measure performance and create benchmarks, and how to optimize performance.

Chapter 12, Evaluating, Recommending, and Presenting Your Solutions, discusses how to evaluate and choose the best-suited alternative among various architectures and how to present the recommended architecture effectively.

To get the most out of this book

It is expected that you have knowledge of Core Java and Maven to get the most out of the book. Basic knowledge of Apache Spark is desirable for *Chapter 5, Architecting a Batch Processing Pipeline*. Basic knowledge of Kafka is desirable for *Chapter 6, Architecting a Real-Time Processing Pipeline*. Also, basic knowledge of MongoDB is good to have to understand the implementation of Chapters 6, 9, and 10.

Software/hardware covered in the book	Operating system requirements
Java SDK 8 or 11	Windows, macOS, or Linux
Apache Maven 3.6 or above	Windows, macOS, or Linux
IntelliJ IDEA Community Edition	Windows, macOS, or Linux
Apache Spark 3.0 or above	Windows, macOS, or Linux
AWS S3, Lambda, EMR, ECR, API Gateway	AWS Cloud
Docker Desktop	Windows, macOS, or Linux
Minikube v1.23.2	Windows, macOS, or Linux
PostgreSQL 14.0	Windows, macOS, or Linux
MongoDB Atlas	AWS cloud
Apache Kafka 2.8.2	Windows, macOS, or Linux
Apache NIFI 1.12.0	Windows, macOS, or Linux
DataHub	Docker/Kubernetes
PostMan	Windows, macOS, or Linux
GraphQL playground 1.8.10	Windows, macOS, or Linux
JMeter 5.5	Windows, macOS, or Linux

You can set up your local environment by ensuring the Java SDK, Maven, and IntelliJ IDEA Community Edition are installed. You can use the following links for installation:

- JDK installation guide: `https://docs.oracle.com/en/java/javase/11/install/overview-jdk-installation.html#GUID-8677A77F-231A-40F7-98B9-1FD0B48C346A`

- Maven installation guide: `https://maven.apache.org/install.html`

- IntelliJ IDEA installation guide: `https://www.jetbrains.com/help/idea/installation-guide.html`

If you are using the digital version of this book, we advise you to type the code yourself or access the code from the book's GitHub repository (a link is available in the next section). Doing so will help you avoid any potential errors related to the copying and pasting of code.

Download the example code files

You can download the example code files for this book from GitHub at `https://github.com/PacktPublishing/Scalable-Data-Architecture-with-Java`. If there's an update to the code, it will be updated in the GitHub repository.

We also have other code bundles from our rich catalog of books and videos available at https://github.com/PacktPublishing/. Check them out!

Download the color images

We also provide a PDF file that has color images of the screenshots and diagrams used in this book. You can download it here: https://packt.link/feLcH.

Conventions used

There are a number of text conventions used throughout this book.

Code in text: Indicates code words in text, database table names, folder names, filenames, file extensions, pathnames, dummy URLs, user input, and Twitter handles. Here is an example: "So, the KStream bean is created as an instance of KStream<String, String>."

```
A block of code is set as follows:
public interface Transformer<K, V, R> {
    void init(ProcessorContext var1);
    R transform(K var1, V var2);
    void close();
}
```

Any command-line input or output is written as follows:

```
bin/connect-standalone.sh config/connect-standalone.properties
connect-riskcalc-mongodb-sink.properties
```

Bold: Indicates a new term, an important word, or words that you see onscreen. For instance, words in menus or dialog boxes appear in bold. Here is an example: "Here, click the **Build a Database** button to create a new database instance."

> **Tips or important notes**
> Appear like this.

Get in touch

Feedback from our readers is always welcome.

General feedback: If you have questions about any aspect of this book, email us at customercare@packtpub.com and mention the book title in the subject of your message.

Errata: Although we have taken every care to ensure the accuracy of our content, mistakes do happen. If you have found a mistake in this book, we would be grateful if you would report this to us. Please visit www.packtpub.com/support/errata and fill in the form.

Piracy: If you come across any illegal copies of our works in any form on the internet, we would be grateful if you would provide us with the location address or website name. Please contact us at copyright@packt.com with a link to the material.

If you are interested in becoming an author: If there is a topic that you have expertise in and you are interested in either writing or contributing to a book, please visit authors.packtpub.com.

Share Your Thoughts

Once you've read *Scalable Data Architecture with Java*, we'd love to hear your thoughts! Scan the QR code below to go straight to the Amazon review page for this book and share your feedback.

https://packt.link/r/1-801-07308-2

Your review is important to us and the tech community and will help us make sure we're delivering excellent quality content.

Section 1 – Foundation of Data Systems

In this section, you will be introduced to various kinds of data engineering problems and the role of a data architect in solving the problems. You will also learn the basics of data format, storage, databases, and data platforms needed to architect a solution.

This section comprises the following chapters:

Basics of Modern Data Architecture

1

With the advent of the 21st century, due to more and more internet usage and more powerful data insight tools and technologies emerging, there has been a data explosion, and data has become the new gold. This has implied an increased demand for useful and actionable data, as well as the need for quality data engineering solutions. However, architecting and building scalable, reliable, and secure data engineering solutions is often complicated and challenging.

A poorly architected solution often fails to meet the needs of the business. Either the data quality is poor, it fails to meet the SLAs, or it's not sustainable or scalable as the data grows in production. To help data engineers and architects build better solutions, every year, dozens of open source and preoperatory tools get released. Even a well-designed solution sometimes fails because of a poor choice or implementation of the tools.

This book discusses various architectural patterns, tools, and technologies with step-by-step hands-on explanations to help an architect choose the most suitable solution and technology stack to solve a data engineering problem. Specifically, it focuses on tips and tricks to make architectural decisions easier. It also covers other essential skills that a data architect requires such as data governance, data security, performance engineering, and effective architectural presentation to customers or upper management.

In this chapter, we will explore the landscape of data engineering and the basic features of data in modern business ecosystems. We will cover various categories of modern data engineering problems that a data architect tries to solve. Then, we will learn about the roles and responsibilities of a Java data architect. We will also discuss the challenges that a data architect faces while designing a data engineering solution. Finally, we will provide an overview of the techniques and tools that we'll discuss in this book and how they will help an aspiring data architect do their job more efficiently and be more productive.

In this chapter, we're going to cover the following main topics:

- Exploring the landscape of data engineering

- Responsibilities and challenges of a Java data architect
- Techniques to mitigate those challenges

Exploring the landscape of data engineering

In this section, you will learn what data engineering is and why it is needed. You will also learn about the various categories of data engineering problems and some real-world scenarios where they are found. It is important to understand the varied nature of data engineering problems before you learn how to architect solutions for such real-world problems.

What is data engineering?

By definition, **data engineering** is the branch of software engineering that specializes in collecting, analyzing, transforming, and storing data in a usable and actionable form.

With the growth of social platforms, search engines, and online marketplaces, there has been an exponential increase in the rate of data generation. In 2020 alone, around 2,500 petabytes of data was generated by humans each day. It is estimated that this figure will go up to 468 exabytes per day by 2025. The high volume and availability of data have enabled rapid technological development in AI and data analytics. This has led businesses, corporations, and governments to gather insights like never before to give customers a better experience of their services.

However, raw data usually is seldom used. As a result, there is an increased demand for creating usable data, which is secure and reliable. Data engineering revolves around creating scalable solutions to collect the raw data and then analyze, validate, transform, and store it in a usable and actionable format. Optionally, in certain scenarios and organizations, in modern data engineering, businesses expect usable and actionable data to be published as a service.

Before we dive deeper, let's explore a few practical use cases of data engineering:

- **Use case 1**: **American Express** (**Amex**) is a leading credit card provider, but it has a requirement to group customers with similar spending behavior together. This ensures that Amex can generate personalized offers and discounts for targeted customers. To do this, Amex needs to run a clustering algorithm on the data. However, the data is collected from various sources. A few data flows from MobileApp, a few flows from different Salesforce organizations such as sales and marketing, and a few data flows from logs and JSON events will be required. This data is known as raw data, and it can contain junk characters, missing fields, special characters, and sometimes unstructured data such as log files. Here, the data engineering team ingests that data from different sources, cleans it, transforms it, and stores it in a usable structured format. This ensures that the application that performs clustering can run on clean and sorted data.

- **Use case 2**: A health insurance provider receives data from multiple sources. This data comes from various consumer-facing applications, third-party vendors, Google Analytics, other marketing platforms, and mainframe batch jobs. However, the company wants a single data repository to be created that can serve different teams as the source of clean and sorted data. Such a requirement can be implemented with the help of data engineering.

Now that we understand data engineering, let's look at a few of its basic concepts. We will start by looking at the dimensions of data.

Dimensions of data

Any discussion on data engineering is incomplete without talking about the dimensions of data. The dimensions of data are some basic characteristics by which the nature of data can be analyzed. The starting point of data engineering is analyzing and understanding the data.

To successfully analyze and build a data-oriented solution, the four *V*s of modern data analysis are very important. These can be seen in the following diagram:

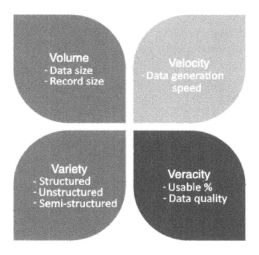

Figure 1.1 – Dimensions of data

Let's take a look at each of these *V*s in detail:

- **Volume**: This refers to the size of data. The size of the data can be as small as a few bytes to as big as a few hundred petabytes. Volume analysis usually involves understanding the size of the whole dataset or the size of a single data record or event. Understanding the size is essential in choosing the type of technologies and infrastructure sizing decisions to process and store the data.

- **Velocity**: This refers to the speed at which data is getting generated. High-velocity data requires distributed processing. Analyzing the speed of data generation is especially critical for scenarios where businesses require usable data to be made available in real-time or near-real-time.

- **Variety**: This refers to the various variations in the format in which the data source can generate the data. Usually, they can be one of the three following types:

  - **Structured**: Structured data is where the number of columns, their data types, and their positions are fixed. All classical datasets that fit neatly in the relational data model are perfect examples of structured data.

  - **Unstructured**: These datasets don't conform to a specific structure. Each record in such a dataset can have any number of columns in any arbitrary format. Examples include audio and video files.

  - **Semi-structured**: Semi-structured data has a structure, but the order of the columns and the presence of a column in each record is optional. A classical example of such a dataset is any hierarchical data source, such as a `.json` or a `.xml` file.

- **Veracity**: This refers to the trustworthiness of the data. In simple terms, it is related to the quality of the data. Analyzing the noise of data is as important as analyzing any other aspect of the data. This is because this analysis helps create a robust processing rule that ultimately determines how successful a data engineering solution is. Many well-engineered and designed data engineering solutions fail in production due to a lack of understanding about the quality and noise of the source data.

Now that we have a fair idea of the characteristics by which the nature of data can be analyzed, let's understand how they play a vital role in different types of data engineering problems.

Types of data engineering problems

Broadly speaking, the kinds of problems that data engineers solve can be classified into two basic types:

- **Processing problems**
- **Publishing problems**

Let's take a look at these problems in more detail.

Processing problems

The problems that are related to collecting raw data or events, processing them, and storing them in a usable or actionable data format are broadly categorized as processing problems. Typical use cases can be a data ingestion problem such as **Extract, Transform, Load** (ETL) or a data analytics problem such as generating a year-on-year report.

Again, processing problems can be divided into three major categories, as follows:

- Batch processing
- Real-time processing
- Near real-time processing

This can be seen in the following diagram:

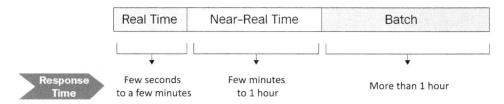

Figure 1.2 – Categories of processing problems

Let's take a look at each one of these categories in detail.

Batch processing

If the SLA of processing is more than 1 hour (for example, if the processing needs to be done once in 2 hours, once daily, once weekly, or once biweekly), then such a problem is called a batch processing problem. This is because, when a system processes data at a longer time interval, it usually processes a batch of data records and not a single record/event. Hence, such processing is called **batch processing**:

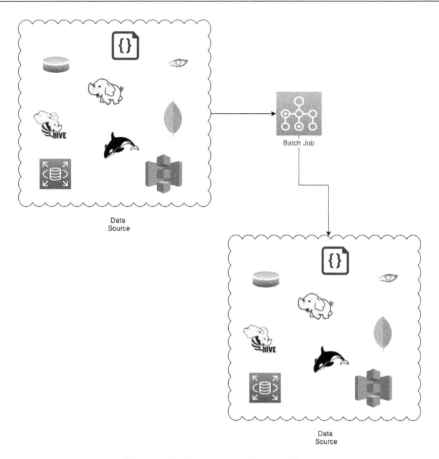

Figure 1.3 – Batch processing problem

Usually, a batch processing solution depends on the volume of data. If the data volume is more than tens of terabytes, usually, it needs to be processed as big data. Also, since big data processes are schedule-driven, a workflow manager or schedular needs to run its jobs. We will discuss batch processing in more detail later in this book.

Real-time processing

A **real-time processing** problem is a use case where raw data/events are to be processed on the fly and the response or the processing outcome should be available within seconds, or at most within 2 to 5 minutes.

As shown in the following diagram, a real-time process receives data in the form of an event stream and immediately processes it. Then, it either sends the processed event to a sink or to another stream of events to be processed further. Since this kind of processing happens on a stream of events, this is known as real-time stream processing:

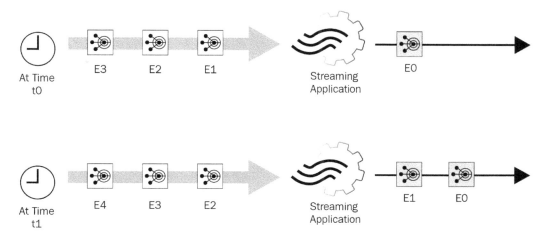

Figure 1.4 – Real-time stream processing

As shown in *Figure 1.4*, event E0 gets processed and sent out by the streaming application, while events E1, E2 and E3 are waiting to be processed in the queue. At t1, event E1 also gets processed, showing continuous processing of events by streaming application

An event can generate at any time (24/7), which creates a new kind of problem. If the producer application of an event directly sends the event to a consumer, there is a chance of event loss, unless the consumer application is running 24/7. Even bringing down the consumer application for maintenance or upgrades isn't possible, which means there should be zero downtime for the consumer application. However, any application with zero downtime is not realistic. Such a model of communication between applications is called **point-to-point** communication.

Another challenge in point-to-point communication for real-time problems is the speed of processing as this should be always equal to or greater than that of a producer. Otherwise, there will be a loss of events or a possible memory overrun of the consumer. So, instead of directly sending events to the consumer application, they are sent asynchronously to an **Event Bus** or a **Message Bus**. An Event Bus is a high availability container that can hold events such as a queue or a topic. This pattern of sending and receiving data asynchronously by introducing a high availability Event Bus in between is called the **Pub-Sub framework**.

The following are some important terms related to real-time processing problems:

- **Events**: This can be defined as a data packet generated as a result of an action, a trigger, or an occurrence. They are also popularly known as **messages** in the Pub-Sub framework.

- **Producer**: A system or application that produces and sends events to a Message Bus is called a **publisher** or a **producer**.

- **Consumer**: A system or application that consumes events from a Message Bus to process is called a **consumer** or a **subscriber**.

- **Queue**: This has a single producer and a single consumer. Once a message/event is consumed by a consumer, that event is removed from the queue. As an analogy, it's like an SMS or an email sent to you by one of your friends.

- **Topic**: Unlike a queue, a topic can have multiple consumers and producers. It's a broadcasting channel. As an analogy, it's like a TV channel such as HBO, where multiple producers are hosting their show, and if you have subscribed to that channel, you will be able to watch any of those shows.

A real-world example of a real-time problem is credit card fraud detection, where you might have experienced an automated confirmation call to verify the authenticity of a transaction from your bank, if any transaction seems suspicious while being executed.

Near-real-time processing

Near-real-time processing, as its name suggests, is a problem whose response or processing time doesn't need to be as fast as real time but should be less than 1 hour. One of the features of near-real-time processing is that it processes events in micro batches. For example, a near-real-time process may process data in a batch interval of every 5 minutes, a batch size of every 100 records, or a combination of both (whichever condition is satisfied first).

At time tx, all events (E1, E2 and E3) that are generated between t0 and tx are processed together by near real-time processing job. Similarly all events (E4, E5 and E6) between time tx and tn are processed together.

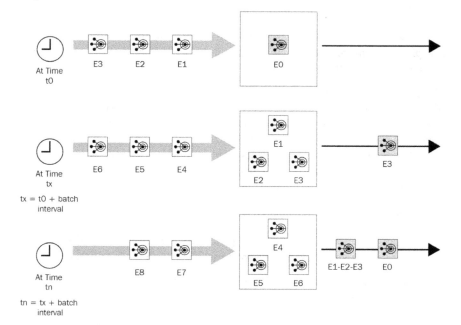

Figure 1.5 – Near-real-time processing

Typical near-real-time use cases are recommendation problems such as product recommendations for services such as Amazon or video recommendations for services such as YouTube and Netflix.

Publishing problems

Publishing problems deal with publishing the processed data to different businesses and teams so that data is easily available with proper security and data governance. Since the main goal of the publishing problem is to expose the data to a downstream system or an external application, having extremely robust data security and governance is essential.

Usually, in modern data architectures, data is published in one of three ways:

- Sorted data repositories
- Web services
- Visualizations

Let's take a closer look at each.

Sorted data repositories

Sorted data repositories is a common term used for various kinds of repositories that are used to store processed data. This is usable and actionable data and can be directly queried by businesses, analytics teams, and other downstream applications for their use cases. They are broadly divided into three types:

- Data warehouse
- Data lake
- Data hub

A **data warehouse** is a central repository of integrated and structured data that's mainly used for reporting, data analysis, and **Business Intelligence** (**BI**). A **data lake** consists of structured and unstructured data, which is mainly used for data preparation, reporting, advanced analytics, data science, and **Machine Learning** (**ML**). A **data hub** is the central repository of trusted, governed, and shared data, which enables seamless data sharing between diverse endpoints and connects business applications to analytic structures such as data warehouses and data lakes.

Web services

Another publishing pattern is where data is published as a service, popularly known as **Data as a Service**. This data publishing pattern has many advantages as it enables security, immutability, and governance by design. Nowadays, as cloud technologies and GraphQL are becoming popular, Data-as-a-Service is getting a lot of traction in the industry.

The two popular mechanisms of publishing Data as a Service are as follows:

- REST
- GraphQL

We will discuss these techniques in detail later in this book.

Visualization

There's a popular saying: *A picture is worth a thousand words*. Visualization is a technique by which reports, analytics, and statistics about the data are captured visually in graphs and charts.

Visualization is helpful for businesses and leadership to understand, analyze, and get an overview of the data flowing in their business. This helps a lot in decision-making and business planning.

A few of the most common and popular visualization tools are as follows:

- **Tableau** is a proprietary data visualization tool. This tool comes with multiple source connectors to import data into it and create easy fast visualization using drag-and-drop visualization components such as graphs and charts. You can find out more about this product at https://www.tableau.com/.
- **Microsoft Power BI** is a proprietary tool from Microsoft that allows you to collect data from various data sources to connect and create powerful dashboards and visualizations for BI. While both Tableau and Power BI offer data visualization and BI, Tableau is more suited for seasoned data analysts, while Power BI is useful for non-technical or inexperienced users. Also, Tableau works better with huge volumes of data compared to Power BI. You can find out more about this product at https://powerbi.microsoft.com/.
- **Elasticsearch-Kibana** is an open source tool whose source code is open source and has free versions for on-premise installations and paid subscriptions for cloud installation. This tool helps you ingest data from any data source into Elasticsearch and create visualizations and dashboards using Kibana. Elasticsearch is a powerful text-based **Lucene** search engine that not only stores the data but enables various kinds of data aggregation and analysis (including ML analysis). Kibana is a dashboarding tool that works together with Elasticsearch to create very powerful and useful visualizations. You can find out more about these products at https://www.elastic.co/elastic-stack/.

> **Important note**
> A Lucene index is a full-text inverse index. This index is extremely powerful and fast for text-based searches and is the core indexing technology behind most search engines. A Lucene index takes all the documents, splits them into words or tokens, and then creates an index for each word.

- **Apache Superset** is a completely open source data visualization tool (developed by Airbnb). It is a powerful dashboarding tool and is completely free, but its data source connector support is limited, mostly to SQL databases. A few interesting features are its built-in role-based data access, an API for customization, and extendibility to support new visualization plugins. You can find out more about this product at `https://superset.apache.org/`.

While we have briefly discussed a few of the visualization tools available in the market, there are many visualizations and competitive alternatives available. Discussing data visualization in more depth is beyond the scope of this book.

So far, we have provided an overview of data engineering and the various types of data engineering problems. In the next section, we will explore what role a Java data architect plays in the data engineering landscape.

Responsibilities and challenges of a Java data architect

Data architects are senior technical leaders who map business requirements to technical requirements, envision technical solutions to solve business problems, and establish data standards and principles. Data architects play a unique role, where they understand both the business and technology. They are like the *Janus* of business and technology, where on one hand they can look, understand, and communicate with the business, and on the other, they do the same with technology. Data architects create processes that are used to plan, specify, enable, create, acquire, maintain, use, archive, retrieve, control, and purge data. According to DAMMA's data management body of knowledge, *a data architect provides a standard common business vocabulary, expresses strategic requirements, outlines high-level integrated designs to meet those requirements, and aligns with the enterprise strategy and related business architecture.*

The following diagram shows the cross-cutting concerns that a data architect handles:

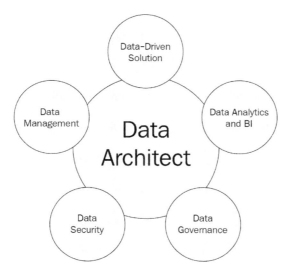

Figure 1.6 – Cross-cutting concerns of a data architect

The typical responsibilities of a Java data architect are as follows:

- Interpreting business requirements into technical specifications, which includes data storage and integration patterns, databases, platforms, streams, transformations, and the technology stack

- Establishing the architectural framework, standards, and principles

- Developing and designing reference architectures that are used as patterns that can be followed by others to create and improve data systems

- Defining data flows and their governance principles

- Recommending the most suitable solutions, along with their technology stacks, while considering scalability, performance, resource availability, and cost

- Coordinating and collaborating with multiple departments, stakeholders, partners, and external vendors

In the real world, a data architect is supposed to play a combination of three disparate roles, as shown in the following diagram:

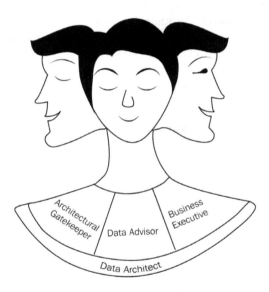

Figure 1.7 – Multifaced role of a data architect

Let's look at these three architectural roles in more detail:

- **Data architectural gatekeeper**: An architectural gatekeeper is a person or a role that ensures the data model is following the necessary standards and that the architecture is following the proper architectural principles. They look for any gaps in terms of the solution or business expectations. Here, a data architect takes a negative role in finding faults or gaps in the product or solution design and delivery (including a lack of or any gap in best practices in the data model, architecture, implementation techniques, testing procedures, **continuous integration/continuous delivery (CI/CD)** efforts, or business expectations).

- **Data advisor**: A data advisor is a data architect that focuses more on finding solutions rather than finding a problem. A data advisor highlights issues, but more importantly, they show an opportunity or propose a solution for them. A data advisor should understand the technical as well as the business aspect of a problem and solution and should be able to advise to improve the solution.

- **Business executive**: Apart from the technical roles that a data architect plays, the data architect needs to play an executive role as well. As stated earlier, the data architect is like the Janus of business and technology, so they are expected to be a great communicator and sales executive who can sell their idea or solution (that is technical) to nontechnical folks. Often, a data architect needs to present elevator speeches to higher leadership to show opportunities and convince them of a solution for business problems. To be successful in this role, a data architect must think like a business executive – *What is the ROI? Or what is there for me in it? How much can we save in terms of time and money with this solution or opportunity?* Also, a data architect should be concise and articulate in presenting their idea so that it creates immediate interest among the listeners (mostly business executives, clients, or investors).

Let's understand the difference between a data architect and data engineer.

Data architect versus data engineer

The data architect and data engineer are related roles. A data architect visualizes, conceptualizes, and creates the blueprint of the data engineering solution and framework, while the data engineer takes the blueprint and implements the solution.

Data architects are responsible for putting data chaos in order, generated by enormous piles of business data. Each data analytics or data science team requires a data architect who can visualize and design the data framework to create clean, analyzed, managed, formatted, and secure data. This framework can be utilized further by data engineers, data analysts, and data scientists for their work.

Challenges of a data architect

Data architects face a lot of challenges in their day-to-day work. We will be focusing on the main challenges that a data architect faces on a day-to-day basis:

- Choosing the right architectural pattern
- Choosing the best-fit technology stack
- Lack of actionable data governance
- Recommending and communicating effectively to leadership

Let's take a closer look.

Choosing the right architectural pattern

A single data engineering problem can be solved in many ways. However, with the ever-evolving expectations of customers and the evolution of new technologies, choosing the correct architectural pattern has become more challenging. What is more interesting is that with the changing technological landscape, the need for agility and extensibility in architecture has increased many folds to avoid unnecessary costs and sustainability of architecture over time.

Choosing the best-fit technology stack

One of the complex problems that a data architect needs to figure out is the technology stack. Even when you have created a very well-architected solution, whether your solution will fly or flop will depend on the technology stack you are choosing and how you are planning to use it. As more and more tools, technologies, databases, and frameworks are developed, a big challenge remains for data architects to choose an optimum tech stack that can help create a scalable, reliable, and robust solution. Often, a data architect needs to take into account other non-technical factors as well, such as the future growth prediction of the tool, the market availability of skilled resources for those tools, vendor lock-in, cost, and community support options.

Lack of actionable data governance

Data governance is a buzzword in data businesses, but what does it mean? Governance is a broad area that includes both workflows and toolsets to govern data. If either the tools or the workflow process has limitations or is not present, then data governance is incomplete. When we talk about actionable governance, we mean the following elements:

- Integrating data governance with all data engineering systems to maintain standard metadata, including traceability of events and logs for a standard timeline

- Integrating data governance concerning all the security policies and standards

- Role-based and user-based access management policies on all data elements and systems

- Adherence to defined metrics that are tracked continually

- Integrating data governance and the data architecture

Data governance should always be aligned with strategic and organizational goals.

Recommending and communicating effectively to leadership

Creating an optimal architecture and the correct set of tools is a challenging task, but it never is enough, unless and until they are not put into practice. One of the hats that a data architect often needs to wear is that of a sales executive who needs to sell their solution to the business executive or upper leadership. These are not usually technical people and they don't have a lot of time. Data architects, most of whom have strong technical backgrounds, face the daunting task of communicating and selling their idea to these people. To convince them about the opportunity and the idea, a data architect needs to back them up with proper decision metrics and information that can align that opportunity to the broader business goals of the organization.

So far, we have seen the role of a data architect and the common problems that they face. In the next section, we will provide an overview of how a data architect mitigates those challenges on a day-to-day basis.

Techniques to mitigate those challenges

In this section, we will discuss how a data architect can mitigate the aforementioned challenges. To understand the mitigation plan, it is important to understand what the life cycle of a data architecture looks like and how a data architect contributes to it. The following diagram shows the life cycle of a data architecture:

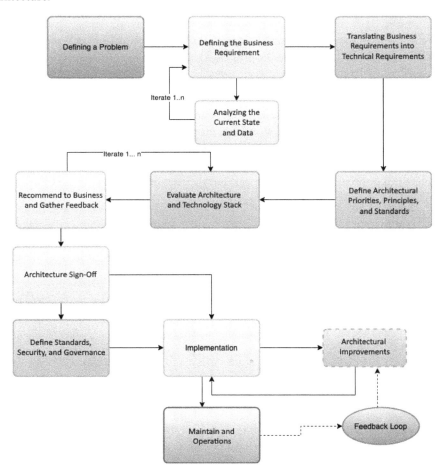

Figure 1.8 – Life cycle of a data architecture

The data architecture starts with defining the problem that the business is facing. Here, this is mainly identified or reported by business teams or customers. Then, the data architects work closely with the business to define the business requirements. However, in a data engineering landscape, that is not enough. In a lot of cases, there are hidden requirements or anomalies. To mitigate such problems, business analysts team up with data architects to analyze data and the current state of the system, including any existing solution, the current cost, or loss of revenue due to the problem and infrastructure where data resides. This helps refine the business requirements. Once the business requirements are more or less frozen, the data architects map the business requirements to the technical requirements.

Then, the data architect defines the standards and principles of the architecture and determines the priorities of the architecture based on the business need and budget. After that, the data architect creates the most suitable architectures, along with their proposed technology stack. In this phase, the data architects closely work with the data engineers to implement **proof of concept** (**POCs**) and evaluate the proposed solution in terms of feasibility, scalability, and performance.

Finally, the architects recommend solutions based on the evaluation results and architectural priorities defined earlier. The data architects present the proposed solutions to the business. Based on priorities such as cost, timeline, operational cost, and resource availability, feedback is received from the business and clients. It takes a few iterations to solidify and get an agreement on the architecture.

Once an agreement has been reached, the solution is implemented. Based on the implementation challenges and particular use cases, the architecture may or may not be revised or tweaked a little. Once an architecture is implemented and goes to production, it enters the maintenance and operations phase. During maintenance and operations, sometimes, feedback is provided, which might result in a few architectural improvements and changes, but they are often seldom if the solution is well-architected in the first place.

In the preceding diagram, the blue boxes indicate major involvement from a customer, a green box indicates major involvement from a data architect, a yellow box means a data architect equally shares involvement with another stakeholder, and a gray box means the data architect has the least involvement in that scenario.

Now that we have understood the life cycle of the data architecture and a data architect's role in various phases, we will focus on how to mitigate those challenges that are faced by a data architect. This book covers how to mitigate those challenges in the following way:

- Understanding the business data, its characteristics, and storage options:

  - Data and its characteristics were covered earlier in this chapter; it will also be covered partly in *Chapter 2, Data Storage and Databases*

  - Storage options will also be discussed in *Chapter 2, Data Storage and Databases*

- Analyzing and defining the business problem:

  - Understanding the various kinds of data engineering problems (covered in this chapter)

- We have provided a step-by-step analysis of how an architect should analyze a business problem, classify, and define it in *Chapter 4, ETL Data Load – A Batch-Based Solution to Ingest Data in a Data Warehouse*, *Chapter 5, Architecting a Batch Processing Pipeline*, and *Chapter 6, Architecting a Real-Time Processing Pipeline*

- The challenge of choosing the right architecture. To choose the right architectural pattern, we should be aware of the following:

  - The types of data engineering problems and the dimensions of data (we discussed this in this chapter)

  - The different types of data and various data storage available (*Chapter 2, Data Storage and Databases*)

  - How to model and design different kinds of data while storing it in a database (*Chapter 2, Data Storage and Databases*)

  - Understanding various architectural patterns for data processing problems (*Chapter 7, Core Architectural Design Patterns*)

  - Understanding the architectural patterns of publishing the data (*Section 3, Enabling Data as a Service*)

- The challenge of choosing the best-fit technology stack and data platform. To choose the correct set of tools, we need to know how to use a tool and when to use what tools we have:

  - How to choose the correct database will be discussed in *Chapter 2, Data Storage and Databases*

  - How to choose the correct platform will be discussed in *Chapter 3, Identifying the Right Data Platform*

  - A step-by-step hands-on guide to using different tools in batch processing will be covered in *Chapter 4, ETL Data Load – A Batch-Based Solution to Ingest Data in a Data Warehouse*, and *Chapter 5, Architecting a Batch Processing Pipeline*

  - A step-by-step guide to architecting real-time stream processing and choosing the correct tools will be covered in *Chapter 6, Architecting a Real-Time Processing Pipeline*

  - The different tools and technologies used in data publishing will be discussed in *Chapter 9, Exposing MongoDB Data as a Service*, and *Chapter 10, Federated and Scalable DaaS with GraphQL*

- The challenge of creating a design for scalability and performance will be covered in *Chapter 11, Measuring Performance and Benchmarking Your Applications*. Here, we will discuss the following:

  - Performance engineering basics

  - The publishing performance benchmark

- Performance optimization and tuning

- The challenge of a lack of data governance. Various data governance and security principles and tools will be discussed in *Chapter 8, Enabling Data Security and Governance*.

- The challenge of evaluating architectural solutions and recommending them to leadership. In the final chapter of this book (*Chapter 12, Evaluating, Recommending, and Presenting Your Solution*), we will use the various concepts that we have learned throughout this book to create actionable data metrics and determine the most optimized solution. Finally, we will discuss techniques that an architect can apply to effectively communicate with business stakeholders, executive leadership, and investors.

In this section, we discussed how this book can help an architect overcome the various challenges they will face and make them more effective in their role. Now, let's summarize this chapter.

Summary

In this chapter, we learned what data engineering is and looked at a few practical examples of data engineering. Then, we covered the basics of data engineering, including the dimensions of data and the kinds of problems that are solved by data engineers. We also provided a high-level overview of various kinds of processing problems and publishing problems in a data engineering landscape. Then, we discussed the roles and responsibilities of a data architect and the kind of challenges they face. We also briefly covered the way this book will guide you to overcome challenges and dilemmas faced by a data architect and help you become a better Java data architect.

Now that you understand the basic landscape of data engineering and what this book will focus on, in the next chapter, we will walk through various data formats, data storage options, and databases and learn how to choose one for the problem at hand.

2
Data Storage and Databases

In the previous chapter, we understood the foundations of modern data engineering and what architects are supposed to do. We also covered how data is growing at an exponential rate. However, to make use of that data, we need to understand how to store it efficiently and effectively.

In this chapter, we will focus on learning how to store data. We will start by learning about various types of data and the various formats of the available data. We will briefly discuss encoding and compression and how well they work with various data types. Then, we will learn about file and object storage and compare these data storage techniques. After that, we will cover the various kinds of databases that are available in modern data engineering. We will briefly discuss the techniques and tricks to choose the correct database for a particular use case. However, choosing the correct database doesn't guarantee a well-built solution. As a data architect, it is important to know how to best design a solution around the database so that we can make the most of the technology we have chosen and have an effective, robust, and scalable data engineering solution in place.

We will end this chapter by discussing how to design data models for different kinds of databases. To help you understand these critical concepts, we will provide hands-on real-world scenarios wherever possible.

In this chapter, we're going to cover the following main topics:

- Understanding data types, formats, and encodings
- Understanding file, block, and object storage
- The data lake, data warehouse, and data mart
- Databases and their types
- Data model design considerations

Understanding data types, formats, and encodings

In this section, you will learn about the various data types and data formats. We will also cover compression and how compression and formats go together. After that, we will briefly discuss data encodings. This section will prepare you to understand these basic features of data, which will be of use when we discuss data storage and databases in the upcoming sections.

Data types

All datasets that are used in modern-day data engineering can be broadly classified into one of three categories, as follows:

- **Structured data**: This is a type of dataset that can easily be mapped to a predefined structure or schema. It usually refers to the relational data model, where each data element can be mapped to a predefined field. In a structured dataset, usually, the number of fields, their data type, and the order of the fields are well defined. The most common example of this is a relational data structure where we model the data structure in terms of an *entity* and a *relationship*. Such a relational data structure can be denoted by crows-foot notation. If you are interested in learning the basics of crows-foot notation, please refer to https://vertabelo.com/blog/crow-s-foot-notation.The following diagram shows an example of structured data in crows-feet notation:

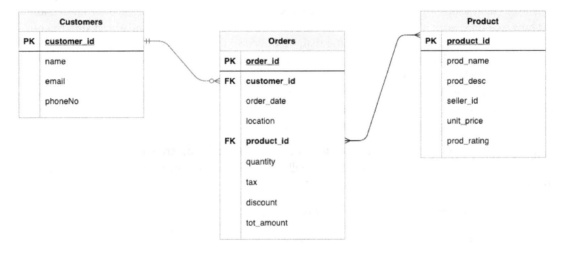

Figure 2.1 – Structured data representation

In the preceding diagram, we can see a system consisting of three structured datasets called Customer, Order, and Product. Each of these datasets has a fixed number of fields and their corresponding data types. We can also see the relationship between the datasets. For example, here, Order is related to Customer via customer_id and Order is related to Product via product_id. Since structured datasets have relationships between them, they are also called relational data models.

- **Unstructured data**: This is a type of data or a dataset that doesn't conform to any predefined data model. Due to the lack of any internal structure, they cannot be stored by any relational data stores such as **Relational Database Management Systems (RDBMSs)**. Also, since there is no schema attached to it, querying and searching is not as easy as in a structured data model.

Around 70% of the data that's generated by systems is unstructured. They can either be generated by humans or by machines:

- **Human-generated unstructured data**: A few examples of human-generated unstructured datasets are media files such as audio and video files, chat, instant messaging, phone call transcriptions, and text messages

- **Machine-generated unstructured data**: A few examples of machine-generated unstructured data are scientific data such as seismic imagery, digital surveillance, and satellite imagery

- **Semi-structured data**: This is a type of dataset that, unlike a relational data model, doesn't contain a tabular structure, but still contains markers or tags to define the hierarchy and field names of the data model. Semi-structured data is hierarchical. Semi-structured data is especially useful for platforms and programming language-agnostic communication between different systems. Before we discuss a few types of semi-structured data, let's look at a real-world example.

Mastercard, Visa, and **American Express (Amex)** are card networks that connect payment processors with issuers. Usually, there are a lot of **business-to-business (B2B)** sales on card networks, where a merchant buys subscription plans to accept a card network, thus increasing the card networks' revenue stream. For example, my dentist accepts only Mastercard and Amex, while Costco stores now only accept Visa all over the US. Each of these huge card networks has many Salesforce orgs or business units such as accounting, sales, and marketing.

Suppose Visa wants to generate a saleability score and the best time to reach a B2B customer. Information gathered from marketing and accounting via Salesforce will be used by a real-time **machine learning (ML)**-based application, which will generate and attach the saleability score and best time to reach the customer. The enriched record, along with this extra information, must flow to the Salesforce org for sales. Salesforce usually uses APEX as a language on the Salesforce cloud (which may be hosted in a different OS), while the AI application that generates the score and best call time is written in Java and Python and sits over an on-premises Kubernetes cluster. For messages to communicate easily between these disparate systems (with different OSs and different languages), we would use a form of semi-structured data (JSON) that is independent of the OS or the language of the different applications involved in this use case.

Now, let's look at a few of the most popular types of semi-structured data:

- **JSON** is the short form of **JavaScript Object Notation**. According to Wikipedia, it is "*an open standard format that uses human-readable text to transmit data objects consisting of attribute-value pairs.*" The following example consists of key-value pairs, where the value can be another JSON as well. This JSON has at least one key-value pair and is a value; this is known as a nested JSON. An array of JSON objects is called a **JSON array**.

The following is an example of a JSON:

```json
{
    "customerId": 47,
    "firstname": "Lilith",
    "lastname": "Wolfgram",
    "address": "324 Spring Ln",
    "city": "Hanover",
    "country": "Micronesia, Federated States of",
    "countryCode": "FO",
    "email": "Lilith.Wolfgram@gmail.com",
    "bills": [
      {
        "billId": 0,
        "billAmount": 4801.98,
        "paymentStatus": false,
        "dueDt": "2020-12-20"
      },
      {
        "billId": 1,
        "billAmount": 668.71,
        "paymentStatus": false,
        "dueDt": "2020-12-27"
      },
      {
        "billId": 2,
        "billAmount": 977.94,
        "paymentStatus": true,
        "dueDt": "2020-11-24"
      }
    ]
}
```

As you can see, there are tags denoting field names such as `customerId`. The `bills` tag's value is an array of JSON objects, so, its value is a JSON array. Also, since `bills` is not a primitive data type but instead another JSON, the preceding JSON is a nested JSON object that shows how JSON has a hierarchical structure.

- **XML** denotes **Extensible Markup Language**. As is evident from the name, it is an open data format that is both human and machine-readable, in which each data value is tagged or marked by a tag that denotes the name of the field. XML is very similar to JSON in passing information between disparate systems in a platform and language-agnostic way. Like JSON, XML is also a hierarchical data structure. XML is the de facto standard for wsdl SOAP APIs. The following is the XML structure for the JSON described earlier:

```
<?xml version="1.0" encoding="UTF-8" ?>
<root>
    <customerId>47</customerId>
    <firstname>Lilith</firstname>
    <lastname>Wolfgram</lastname>
    <address>324 Spring Ln</address>
    <city>Hanover</city>
    <country>Micronesia, Federated States of</country>
    <countryCode>FO</countryCode>
    <email>Lilith.Wolfgram@gmail.com</email>

    <bills>
        <bill>
            <billId>0</billId>
            <billAmount>4801.98</billAmount>
            <paymentStatus>false</paymentStatus>
            <dueDt>2020-12-20</dueDt>
        </bill>
        . . .
    </bills>
</root>
```

The full source code for the preceding code snippet is available on GitHub at https://github.com/PacktPublishing/Scalable-Data-Architecture-with-Java/blob/main/Chapter02/sample.xml.

As you can see, each XML starts with a root tag, and each value is encapsulated by the tag names. We explored various data types in this section, but we need to understand how these types of data are formatted. So, in the next section, we will discuss various data formats.

Data formats

In data engineering, a dataset can be stored in different kinds of file formats. Each format has its pros and cons. However, it is important to know which data format is more suitable for certain types of use cases over others. In this section, we will discuss the various characteristics of data formats, a few popular data formats, and their suitable use cases.

Characteristics of data formats

First, let's review the various characteristics of the data format that makes them different. These characteristics also determine when a particular data type should be selected over others to solve a business problem. The main characteristics of data formats are as follows:

- **Text versus binary**: Any data file is either stored as a text or binary file. While both text and binary files store the data as a series of bits (0s and 1s), the bits in a text file can only represent characters. However, in a binary file, it can either be a character or any custom data. A binary file is a file that contains binary data as a series of sequential bytes in a specific format, which can be viewed or interpreted by a specific kind of application. A binary file will contain all the information required by the reader application to view/edit the data. For example, a `.jpg` file can only be opened by applications such as Photos. Text files, on the other hand, can only contain characters, can be opened in any text editor, and are human-readable. For example, any `.txt` file that can be opened by text editors such as Notepad is a text file.

- **Schema support**: A schema is an outline, diagram, or model that defines the structure of various types of data. The schema stores field-level information such as the *data type*, *max size*, or *default value*. A schema can be associated with the data, which helps with the following:

  - Data validation

  - Data serialization and compression

  - A way to communicate the data to all its consumers to easily understand and interpret it

  Data formats may or may not support schema enforcement. Also, a schema can be included along with the data, or it can be shared separately with the schema registry. The schema registry is a centralized registry of schemas so that different applications can add or remove fields independently, enabling better decoupling. This makes it suitable for schema evolution and validation.

- **Schema evolution**: As a business grows, more columns get added or changes are made to the column data type, which results in the schema changing over time. Even as the schema evolves, it is important to have backward compatibility for your old data. Schema evolution provides a mechanism to update the schema while maintaining backward compatibility. A few data formats, such as **Avro**, support schema evolution, which is helpful in an agile business; as a dataset's schema can change over time.

- **Row versus columnar storage**: To understand row versus columnar storage, let's take a look at the following diagram:

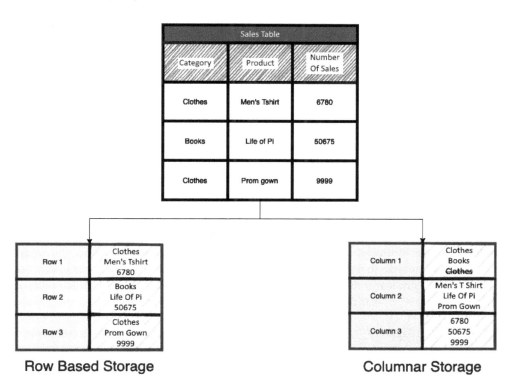

Figure 2.2 – Row versus columnar storage

The preceding diagram shows how the same data will be stored in row-based storage versus column-based storage. This example shows how sales data is stored in a columnar format versus a row-based data storage format. In a row-based format, all data is stored row-wise – that is, the columns of a specific row are stored adjacent to each other. Since the data is stored row-wise, it is ideal for scenarios where it is preferable to read or write data row-wise, such as in **Online Transaction Processing (OLTP)**.

On the other hand, as evident in the preceding diagram, columnar storage stores the values of the same column in an adjacent memory block. So, columns are stored together. Since it stores data in a columnar fashion, it can optimize storage space by storing repetitive column values once and pointers for each row (this is indicated by a striking of the repetitive **Clothes** value in the **Columnar Storage** part of the preceding diagram). This kind of storage is very useful for scenarios where only a subset of columns is read repetitively and doesn't expect transactional writes. Two typical use cases that are widely used are **Online Analytical Processing (OLAP)** and big data processing. Both are mainly used for analytical queries on huge datasets. In big data, the columnar format gives the added advantage of being splittable and enables partition creation, which helps process the data faster.

- **Splittable**: Another important factor is whether the file can be partitioned or split into multiple files. This factor plays a role when the data volume is huge or the velocity is too high, as in the case of big data. Big data files can be stored in a distributed filesystem such as **HDFS** if the underlying data format is splittable. By doing so, processing such partitioned big data becomes much faster.

- **Compression**: Data processing performance often depends on the data's size. Compression reduces the size of the data on disk, which increases network I/O performance (however, it might take more time to decompress it while processing). It also reduces the data packet size when this data flows over the network, and hence the data transfer rates as well. The following table shows a few popular data compression algorithms and their features:

Name	Lossless Compress	Compression Ratio	Splitable	Compression Speed	Decompress Speed
Gzip	Yes	2.7x-3x	No	100 MBps	440 MBps
Snappy	Yes	2x	No	580 MBps	2020 MBps
LZ4	Yes	2.5x	No	800 MBps	4220 MBps
Zstd	Yes	2.8x	Yes	530 MBps	1360 MBps

Table 2.1 – Different compression techniques

- **Companion technologies**: Sometimes, the choice of data format is dependent on a companion technology. For example, in a Hadoop environment, if we are planning to process the data using a Hive MapReduce job, it might be a good idea to use ORC format over Parquet format. But on the other hand, if all our transformations are done using Apache Spark, Parquet may be a better choice.

In this section, we learned about the features and characteristics of various data formats and how they affect the storage and processing of data elements. However, it is important for an architect to be aware of the popular data formats and how they can be used judiciously.

Popular data formats

In this section, we will discuss a few popular data formats that are worth knowing about. You will encounter these when you try to develop a data engineering solution. They are as follows (we covered two popular data formats, JSON and XML, in the *Semi-structured data* section, in the *Data types* subsection):

- **Delimiter Separated Format**: This is a text data format where newline is used as a record delimiter and there can be specific field delimiters based on which type of delimiter-separated file we are dealing with. Two of the most popular delimiter-separated formats are **Comma Separated Value** (CSV) and **Tab Separated Value** (TSV). While in CSV, the field delimiter is a comma, for TSV it is a tab. Optionally, they can have a header record. Although it doesn't support splitting, it provides a very good compression ratio. This format doesn't support null values or schema evolution.

   Due to the simplicity of the format, it is quite popular in batch processing scenarios as well as real-time stream processing. However, the lack of schema evolution, partitioning capabilities, and non-standardized formatting makes its usage limited and not recommended for many use cases.

- **Avro**: This is a row-based data storage format that is known for its serialization capabilities. It stores its data in binary format compactly and efficiently. Avro has great support for schema enforcement and schema evolution. An Avro schema is defined in JSON format. Avro files are not human-readable from a text editor. However, Avro data can be read in a human-readable format such as JSON using `avro-tools-<version>.jar`. The command to convert Avro into a human-readable JSON format is as follows:

   ```
   java -jar ~/avro-tools-1.7.4.jar tojson filename.avro
   ```

   `avro` data is always accompanied by its schema, which can be read using `avro-tools-<version>.jar`, like so:

   ```
   java -jar ~/avro-tools-1.7.4.jar getschema filename.avro
   ```

   If we had a binary Avro file equivalent to that of the JSON described while explaining Semi-structured data in the *Data types* section, the `avro` schema would look as follows:

   ```
   {
     "name": "MyClass",
     "type": "record",
     "namespace": "com.sample.avro",
     "fields": [
       {
   ```

```
      "name": "customerId",
      "type": "int"
    },
    {
      "name": "firstname",
      "type": "string"
    },
    {
      "name": "lastname",
      "type": "string"
    }
    ... ]
}
```

The full source code for the preceding code snippet is available on GitHub at https://github.com/PacktPublishing/Scalable-Data-Architecture-with-Java/blob/main/Chapter02/avroschema.json.

- **Parquet** is an open source column-based data storage that's ideal for analytical loads. It was created by Cloudera in collaboration with Twitter. Parquet is very popular in *big data engineering* because it provides a lot of storage optimization options, as well as provides great columnar compression and optimization. Like Avro, it also supports splittable files and schema evolution. It is quite flexible and has very good support for nested data structures. Parquet gives great read performance; it especially works very well with Apache Spark

- Let's try to understand how a Parquet file is structured. The following diagram shows the Parquet file structure:

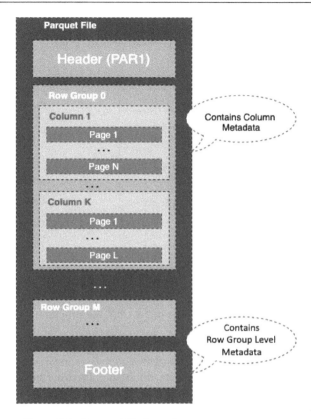

Figure 2.3 – Parquet file format

A Parquet file contains a header and a footer. The header only consists of a marker called **PAR1**, denoting that it's a parquet file. Then, the file is divided into row groups, where each row group denotes a set of rows contained in a chunk of data. Each chunk of data is equal to the block size of the Parquet file (128 MB by default). Each row group contains a chunk for every column. Again, each column chunk consists of one or more pages. Each page in a column consists of n number of rows whose size is less than or equal to a configured page size. Each column chunk stores metadata as well (min/max value, number of nulls, and so on). The footer contains metadata for all row groups, as well as the schema of the data.

- **Optimized Row Columnar** (**ORC**): This is yet another open source file format developed in the Hadoop ecosystem by Hortonworks in collaboration with Facebook. ORC is another form of columnar data storage that supports excellent compression and column optimization. Let's look at the ORC file structure to understand how it is different from the Parquet format. The following diagram shows the structure of an ORC file format:

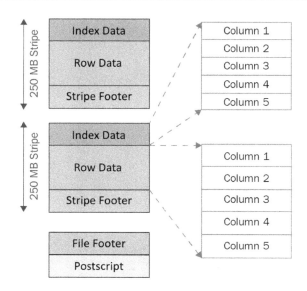

Figure 2.4 – ORC file structure

In ORC format, each file consists of multiple strips. The default stripe size is 250 MB. Each stripe is subdivided into index data, row data, and a stripe footer. The index data contains indexes and the row data consists of actual data, but both of them are stored in columnar format. The stripe footer contains column encodings and their location. The file footer contains information about the list of stripes, the number of rows in each stripe, and the data type of each column. Apart from that, it contains stripe-level statistical information such as min, max, and sum. Finally, the postscript contains information regarding the length of the file's footer, metadata section, and compression-related information.

Let's now understand how to choose from the different data formats.

How to choose between Avro, Parquet, and ORC

To choose the correct data format, we must consider the following:

- **Read or write-intensive query pattern**: For a write-intensive use case, a row-based format works better as appending new records becomes easier. So, Avro would be a better choice for write-intensive use cases. On the other hand, if a read-intensive use case needs to read a subset of columns more frequently, a columnar data format such as Parquet or ORC is a suitable choice.

- **Compression**: This is a very important aspect when choosing a data format because compression reduces both the time and storage required to store or transmit data. For big data use cases, compression involves a huge role. Row-based storage is not suitable for such scenarios. So, for big data analytical use cases, columnar storage such as Parquet or ORC is preferred. Also, more compression is required if transforming/processing big data creates a lot of intermediate reads and writes. In such a scenario, ORC is preferred because it gives a better compression ratio than Parquet. For example, if you are running a MapReduce job or a **Hive Query Language** (**HQL**) query on Hive with the MapReduce engine, ORC will perform better than Parquet.

- **Schema Evolution**: In many data engineering use cases, schemas change frequently over time as new columns get added or dropped as business requirement changes. If there are frequent changes in the data schema and you need backward schema compatibility, Avro is the best choice. Avro supports very advanced schema evolution, compatibility, and versioning while keeping the schema definition simple in JSON format.

- **Nested Columns**: If your use case is suitable for a row-based format, Avro works great with the nested column structure. Otherwise, if the use case is suitable for a columnar data format and you have a lot of nested complex columns, then Parquet is the ideal data format for such use cases.

- **Platform Support**: Finally, the platform or framework plays a very important role. Hive works best with ORC, while Apache Spark and Delta Lake have great support for Parquet. Avro or JSON is often a good choice for Kafka.

In this section, we learned about various data formats such as text, Parquet, Avro, and others. In the next section, we will learn how data (which can be in text, Parquet, or any other format) can be stored using different data storage formats.

Understanding file, block, and object storage

In this section, we will cover the various data storage formats that are essential for an architect who is planning to store their data. Data storage formats organize, keep, and present data in different ways, each with its pros and cons. The available data storage formats are file, block, and object.

File storage organizes and exposes data as a hierarchy of files and folders, whereas block storage divides the data into chunks and stores them in organized, evenly sized volumes. Finally, object storage manages the data in a space-optimized fashion and links it to its associated metadata.

Now, let's dive deeper by discussing their basic concepts, pros and cons, and the use cases where they are applied. Let's begin by discussing the simplest and the oldest of them all: file storage.

File storage

In file-level storage, data is stored as a single piece of information inside a file. This file is given a name, can contain metadata, and resides inside a directory or subdirectory. When you need to find a file, the computer needs to know the absolute path of the file to search and read the file.

The pros and cons of file storage are as follows:

- **Pros**: It's simple, has broad capabilities, and can store anything

- **Cons**: Not ideal for storing huge volumes as there is no option to scale up, only to scale out

A few typical use cases are as follows:

- File storage is ideal for file sharing in offices and other environments for its sheer simplicity; for example, NAS.

- Local archiving. NAS provides excellent support for storing archival data.

- Data protection and security. File-level storage is an old technology, but due to the test of time and a broad variety of uses, its policy, standard, and protection capabilities are all advanced. This makes it a great candidate for data protection use cases.

Let's now take a look at block-level storage.

Block storage

In block-level storage, the data is divided into small chunks of data and assigned unique chunk identifiers. Since the chunk of data is small and has a unique identifier, it can be stored anywhere. Also, a group of data chunks consists of a logical unit called a volume. In block-level storage, you can add a volume of data easily to scale up the infrastructure by adding blocks.

One of the interesting things is how it handles metadata. Unlike a file-based architecture, there are no additional details associated with block storage other than its address. Here, the operating system controls the storage management, which makes it ideal storage for high-performance use cases.

The pros and cons of block-level storage are as follows:

- **Pros**: It provides metadata handling by controlling the OS or database, making it highly performant. You can also easily scale storage up and down.

- **Cons**: It can be expensive. Also, externalizing metadata handling in the application layer means more headaches when managing metadata.

A few typical use cases are as follows:

- **Databases**: Databases usually use block storage. For example, AWS Relational Data Service uses AWS Elastic Block Storage volumes as its storage to store data.

- **Virtualization**: Virtualization software such as **VMware**, **Hyper-V**, and **Oracle VirtualBox** use block storage as their filesystems for the virtual operating system.

- **Cloud-based instances**: Cloud-based instances such as AWS EC2 use block storage (AWS Elastic Block Storage) as their hard disk storage.

- **Email servers**: Microsoft's email server, Exchange, uses block storage as its standard storage system.

Let's look at object-level storage next.

Object storage

Object-level storage stores data in isolated containers called objects, which have unique identifiers and flat structures. This makes data retrieval super easy as you can retrieve an object by using the unique identifier, irrespective of the location it is stored.

The pros and cons of object-level storage are as follows:

- **Pros**: Object storage provides great *metadata flexibility*. For example, you can customize metadata so that the application is associated with an object or you can set the priority of an application to an object. You can pretty much do any customization. This flexibility makes object storage strong and super easy to manage.

  Apart from metadata flexibility, object storage is known for its *accessibility* as it has a REST API to access, which makes it accessible from any platform or language.

  Object storage is extremely *scalable*. Scaling out object architecture is as simple as adding nodes to an existing storage cluster. With the rapid growth of data and the cloud's pay-as-you-go model, this feature has helped object storage become the most sought-after storage for data engineering needs of the present and future.

- **Cons**: With so many positives, there are certain drawbacks to object storage. The biggest and most notable one is that objects can't be modified. However, you can create a newer version of the object. In some use cases such as big data processing, this is a boon instead of a headache.

A few typical use cases are as follows:

- **Big data**: Due to scalability and metadata flexibility, huge volumes, as well as unstructured data, can be stored and read easily from object storage. This makes it suitable for big data storage.

- **Cloud**: Again, due to scalability, object storage is a perfect candidate for cloud systems. Amazon S3 is Amazon's object storage solution and is very popular. Also, customizable metadata helps Amazon S3 objects have a life cycle defined through the AWS console or its SDKs.

- **Web Apps**: Object storage's easy accessibility using a REST API makes it a perfect candidate to be used as a backend for web apps. For example, AWS S3 alone is used as a cheap and quick backend for static websites.

With that, we have covered the various kinds of data storage. In the next section, we will learn how enterprise data (stored in any of the aforementioned storage formats) is organized into different kinds of data repositories, which enables other applications to retrieve, analyze, and query that data.

The data lake, data warehouse, and data mart

To build a data architecture, an architect needs to understand the basic concept and differences between a data lake, data warehouse, and data mart. In this section, we will cover the modern data architectural ecosystem and where the data lake, data warehouse, and data mart fit into that landscape.

The following diagram depicts the landscape of a modern data architecture:

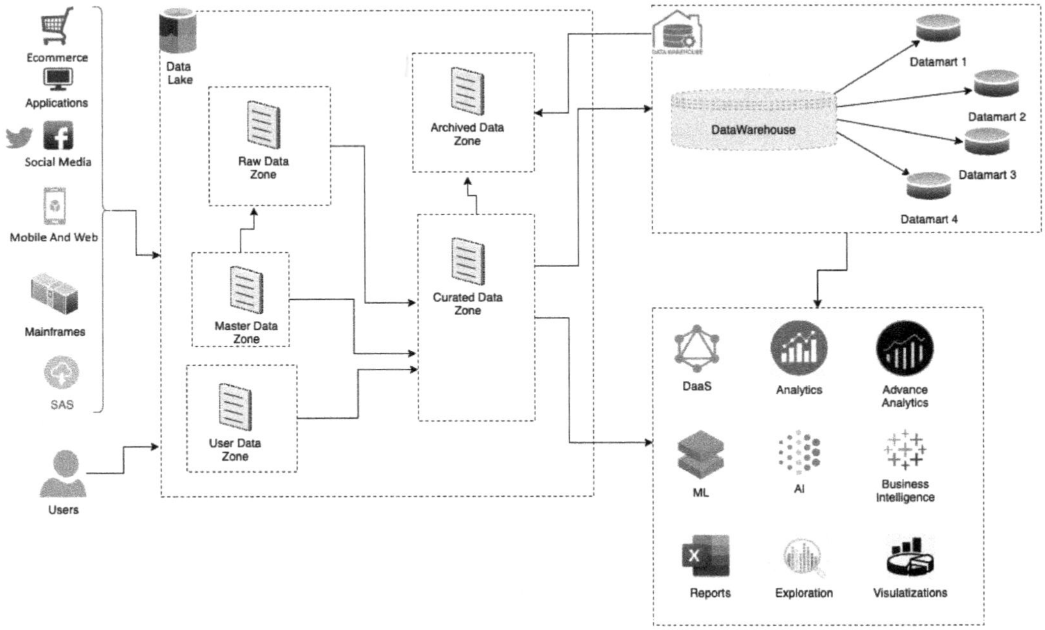

Figure 2.5 – Landscape of a modern data architecture

As we can see, various types of data get ingested into the data lake, where it lands in the raw zone. The data lake consists of structured, semi-structured, and unstructured data ingested directly from data sources. Data lakes have a zone consisting of cleansed, transformed, and sorted datasets that serve various downstream data processing activities such as data analytics, advanced analytics, publishing as Data-as-a-Service, AI, ML, and many more. This is called the **curated zone**. The data lake acts as a source for creating a data warehouse, which is a structured data repository built for a specific line of business.

Data lake

In a modern data architecture, data from various sources is ingested into a data lake. A data lake is a data storage repository that contains structured, semi-structured, and unstructured data. In most cases, the usage of the data in a data lake is not predefined. Usually, once data is ingested and stored in a data lake, various teams use that data for analytics, reports, business intelligence, and other usages.

However, internally, a data lake contains different data zones. The following are the different data zones that are available in a data lake:

- **Raw data zone**: The raw data from various data sources is loaded into this zone. Here, the data that's loaded is in raw form. This data can be unstructured, uncleaned, and unformatted. This is also known as the landing zone.

- **Master data zone**: This data zone usually contains reference data that augments the analytical or transformational activities of data present in the raw zone or curated zone.

- **User data zone**: Sometimes, in certain data lakes, the user can manually drop certain data. They are usually static. This portion of the data lake is called the user data zone.

- **Curated data zone**: This is the data publishing layer of the data lake. It contains cleansed, transformed, and sorted data. The data present in this layer is usually structured. Data may be stored in large flat files, as key-value stores, as data documents, in a star schema, or in a denormalized format. All data governance, data management, and security policies apply to this layer as this is the main consumption layer of the data lake.

- **Archived data zone**: The archive zone consists of data that has been offloaded by other systems such as a data warehouse or the curated zone due to aging. Data in this zone can't usually be modified but can be appended. This kind of data is used for historical analysis or auditing purposes. Usually, a cheaper data storage technology is used to store archived data. Technologies such as Amazon S3 provide more advanced capabilities to progressively move data to cheaper solutions automatically as time progresses using an S3 bucket's life cycle policy.

Let's move on to data warehouses next.

Data warehouse

A data warehouse is a sorted central repository that contains information that has been curated from multiple data sources in a structured user-friendly fashion for data analytics. A good amount of discovery, analysis, planning, and data modeling is required before ingesting the data in a data warehouse. It is highly cleansed, transformed, and structured. As evident from *Figure 2.6*, the data warehouse is built from a data lake in modern data engineering pipelines. While data lakes are usually centralized raw data zones for the enterprise or organization, data warehouses are usually built per business unit or department. Each data warehouse is structured such that it caters to the need of that particular department. A deep dive into data warehouses and their schema types will be discussed in *Chapter 4, ETL Data Load – A Batch-Based Solution to Ingest Data in a Data Warehouse*.

Data marts

Data marts are usually a subset of a data warehouse that focuses on a single line of business. While a data warehouse is typically few 100 GBs to TBs in size, data marts are usually less than 100 GB in size. Data marts provide great read performance as it contains data which is analyzed, designed, and stored for a very specific line of business. For example, from a centralized company data warehouse, there can be a specific data mart for the HR department, one for the finance department, and another for the sales department.

The following table captures the difference between a data lake and a data warehouse:

Characteristics	Data Lake	Data warehouse
Load Pattern	ETL (Extract, Load, and Transform)	ETL (Extract, Transform, and Load)
Type Of Data Stored	Structured, semi-structured and unstructured	Structured
Analysis Pattern	Acquire, analyze, and then determine structure of curated data	Create the structure first and then acquire the data for insights
Data Ingestion Pattern	Batch processing, real-time, Batch processing near real-time processing	Batch processing
Schema Application Time	Schema-on-read i.e., schema is applied while reading the data	Schema-on-write i.e., schema is determined and is available when data is written

Table 2.2 – Data lake versus a data warehouse

So far, we have learned how various data repositories are used and how they enable enterprise data platforms. Data in these repositories can be stored as files or objects, but they can be stored in an organized data collection called a database, which can retrieve, manage, and search data easily. In the next section, we will discuss databases in detail.

Databases and their types

In this section, we will cover the various types of databases that are commonly used to create modern data engineering solutions. We will also try to explore the possible scenario when a specific type of database will be used.

A database is a systematic collection of data or information that's stored in such a way that it can easily be accessed, retrieved, and managed. In modern-day data engineering, primarily, databases can be broadly classified into two categories, as follows:

- **Relational database**: This is a kind of database known for storing structured datasets. Each type of dataset is related to another, and relational databases provide an easy way to establish a relationship between different kinds of datasets. We will discuss relational databases in detail later in this chapter.

- **NoSQL databases** or **non-relational databases**: NoSQL databases are non-relational databases, where data can be stored in some form other than a tabular format. NoSQL supports unstructured, semi-structured, and structured data. No wonder NoSQL stands for *Not only SQL*!

The following diagram depicts the types of databases employed in a modern data engineering context:

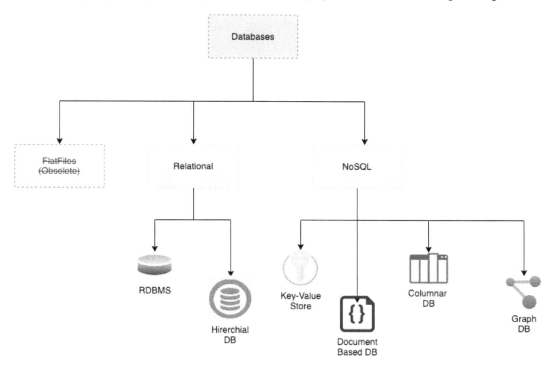

Figure 2.6 – Types of databases

Now, let's discuss the various database types in detail.

Relational database

A relational database, as discussed earlier, stores structural data. Each type of data in a relational database is stored in a container called a database table or, simply, a table. Each table needs to be defined first before data is loaded into that table. The table definition contains the column names or field names, their data type, and their size (optionally). Relational databases are further subdivided into two types: hierarchical databases and RDBMSs.

Hierarchical database

These are databases where data is stored in a tree-like structure. The databases consist of a series of data records. Each record contains a set of fields that are determined by the type of records (this is also called a segment). Each segment can be related to another segment by relationships called *links*. These types of databases are known for *parent-child relationships*. The model is simple but can only support one-to-many relationships. The following diagram shows an example of a hierarchical database model:

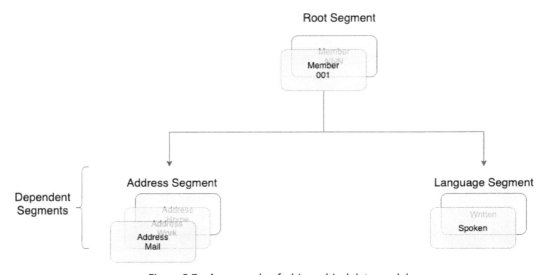

Figure 2.7 – An example of a hierarchical data model

As shown in the preceding diagram, **Member** is the root segment. One record of the **Member** segment contains the ID **001**. There are two child segments to the root segment called **Address** and **Language**. In the **Address Segment** part, we can see three record instances – that is, **Address Mail**, **Address Home**, and **Address Work**. The **Language Segment** part also has instances such as **Spoken** and **Written**.

Examples of hierarchical databases include IBM **Information Management System (IMS)** and RDM Mobile.

RDBMS

RDBMS is a relational database management system that uses SQL as its programming and querying interface. It is the most popular and established kind of database across the industry. Data is stored in tables, which represent specific entities. Tables have a clearly defined set of columns, along with their data type. Each row in a table is called a record. Each table can contain primary keys that uniquely identify a record. Each table supports a variety of indexes. One table can be linked to another table by a foreign key index. RDBMS can support one-to-many and many-to-many relationships. They are very powerful and have been the powerhouses behind most modern applications for many decades.

Examples of RDBMSs include MySQL, Oracle, PostgreSQL, Amazon RDS, and Azure SQL.

When to use: RDBMS is pretty much used everywhere you need multi-row ACID transactions and where you require complex joins. Web applications, employee management systems, and financial organization's online transactions are a few examples of where RDBMS is used.

NoSQL database

NoSQL, as discussed earlier in this section, supports unstructured as well as semi-structured data. This is possible because it supports flexible schema. Also, NoSQL databases store and process data in a distributed manner and hence can scale out infinitely. The usage of distributed computing in NoSQL database architectures helps them support a tremendous volume of data and makes them a great choice for big data processing. The different ways a relational database and a NoSQL database handle scaling can be seen in the following diagram:

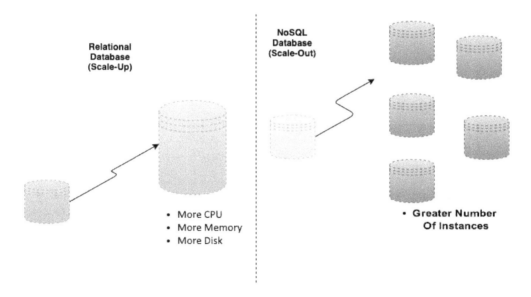

Figure 2.8 – Scale-up versus scale-out

As we can see, a relational database scales up the same instance. However, that creates a limitation of scaling. Also, scaling up is a costlier operation. On the other hand, NoSQL uses commodity hardware, which is cheap, and the architecture is such that to scale it, it needs to be scaled out. This means that NoSQL can scale infinitely and is cheaper to scale.

NoSQL databases can be further categorized into specific kinds of databases. We will discuss each of them briefly while providing examples and usages.

Key-value store

Key-value stores are the simplest kind of NoSQL databases. The data that's stored is in a key and value format. The attribute name is stored in the *key*, while the value of the attribute is stored in the *value*. Here, the key needs to be a string, but the value can be an object of any type. This means that the value can be a JSON, an XML, or some custom serialized object.

A few examples of key-value stores are Redis, Memcached, and RocksDB.

When to use:

- In a microservice or an application. If you need a lookup table that needs to be read fast, then an in-memory key-value store such as Redis and Memcached is a good choice. Again, while Memcached supports concurrent reads, it doesn't support values that are complex like Redis does. Cloud services such as AWS ElastiCache support both of these databases. If you are interested, you can find a more detailed comparison between Redis and Memcached at https://aws. amazon.com/elasticache/redis-vs-memcached/.

- In real-time event stream processing, the state needs to be maintained if the current event processing is dependent on the state of an older event. This kind of real-time processing is called stateful stream processing. In stateful stream processing, RocksDB is a great choice to maintain the state as a key-value pair. Kafka Streams uses RocksDB internally to maintain the state for stateful stream processing.

Next, let's take a look at document-based databases.

Document-based database

Document databases are NoSQL databases that give you an easy way to store and query data from a document. A document is defined as a semi-structured data format such as JSON or XML. Document databases support nested elements as well, such as nested JSON and JSON arrays. Each document in a document database is stored in a key-value pair, where the key is a unique ID for the document and the value is the document that is stored. Document databases support indexing on any field of the document, even if it is a nested field.

A few examples of document-based databases are MongoDB, Apache CouchDB, Azure Cosmos DB, AWS DocumentDB, and ElasticSearch.

When to use:

- When you want to publish curated data from a data lake or a data mart to web applications by using a microservice or REST API. Since web applications run on JavaScript, they can easily parse a JSON document. Storing a well-designed JSON document in a document database such as MongoDB or AWS DocumentDB gives the web application amazing performance.

- If you are receiving feeds from multiple dynamic data feeds, such as social media feeds from Twitter, LinkedIn, and Facebook, and the schema of these feeds is evolving, you must process and publish this data together by extracting certain data points or running some kind of aggregation over them, then Apache CouchDB may be an excellent choice. Simply put, if you are consuming document data and have no control over the inbound schema, a document-based data store is a great choice.

- If your lookup needs are not catered by a key-value store. If the value is a document that has a very complex schema or that the cost of the storage in the key-value store is becoming too high because of the volume of the data, then a document-based database is the next most obvious choice.

- If you are creating a search repository for a business, then you might want to store the data in a search engine storage such as Elasticsearch, a document-based database. It creates a reverse text index (called the Lucene index) while storing the data. This is a special document-based database where each record is stored as a document, along with a unique key. Elasticsearch provides amazing search performance. However, data should only be stored in Elasticsearch if you want to perform a high-performance text-based search over the data or to create some visualization out of the data.

Let's now explore columnar databases.

Columnar database

A columnar database stores data in a columnar format. Columnar databases are created using Big table. According to a paper published by Google that introduced Bigtable, it is *a sparse, distributed, persistent multidimensional sorted map*. At its core, each columnar database is a map. Here, each data record is associated with a key called the row key. These keys are unique and lexicographically sorted. The data that's stored in a columnar database is persisted in a distributed filesystem that provides high availability of the data. Instead of columns, we define column families in a columnar database. Each column family can consist of any number of columns. The columns inside a column family are not fixed for all records and can be added dynamically. This means that in most data records, one or more columns may be empty or non-existent, so this data structure is sparse. This allows you to dynamically add columns to a record. This makes columnar databases a great choice for storing unstructured data. The following diagram tries to capture the essence of a columnar database:

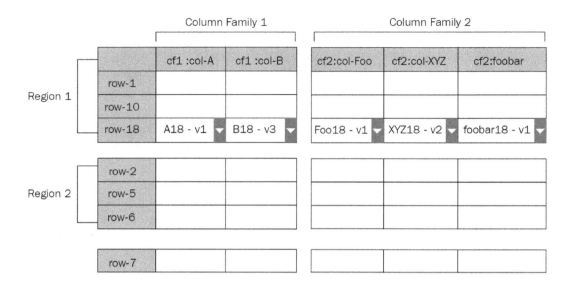

Figure 2.9 – Columnar database structure

As shown in the preceding diagram, the records are divided into regions. One or more regions reside on a node in a distributed filesystem such as HDFS or GFS. Each column family inside a region is stored as a separate file. Again, each column inside a column family can support versioning, which makes columnar storage truly multi-dimensional.

Examples include Apache HBase, Cassandra, and Apache Kudu.

When to use:

- In ad agency and marketing campaigns, a columnar data store is used to store the events of user clicks and user choices in real time. These real-time events are used on the fly to optimize the ads shown to a user or offers to send to a customer.

- Another example is data received from Kafka as a stream of events that are small in size. These need to be stored in HDFS so that they can be analyzed or processed periodically using some form of batch application. Here, a columnar database is preferred since storing the data directly in HDFS or a warehouse such as Hive will create too many small files, which, in turn, will create too much metadata, thus slowing down the Hadoop cluster's overall performance. Columnar storage is written to disk when the region size is reached and is usually placed in sequential files, so they are ideal for this kind of storage.

- They are great databases for massive dynamic spikes in data. For example, they are great for handling massive data surges when sales are on during the holiday season.

Next, let's take a look at the graph database.

Graph database

A graph database is a database where data is stored in a graph structure. Essentially, this means that graph databases not only store the data but also the relationships between it. With the advent of social networking and since the data of every domain has become more connected, there is a growing need to not only query data but query the connections between the data. Also, in social networks, it is necessary to explore neighboring data points (for example, LinkedIn needs to explore data adjacency to show whether a person is connected to your profile as a 1st level, 2nd level, or 3rd level connection). Although relational databases can be used to get relationships using joins, a database can store, process, and query connections efficiently only if it natively supports relationships.

Most graph databases use a popular modeling approach called the **property graph model**. Here, data is organized into nodes, relationships, and properties. The following diagram shows an example of data stored using the property graph model:

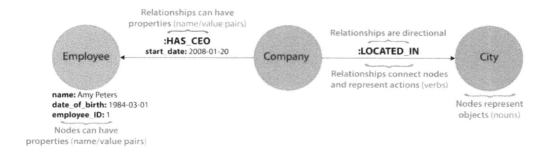

Figure 2.10 – Example of a property graph model

In the property graph model, there are nodes and relationships. **Nodes** are the entities in the model such as **Employee**, **Company**, and **City**. These entities have attributes (key-value pairs), which are called the properties of that entity. In the preceding example, **Employee** has properties such as name, data_of_birth, and employee_ID.

Relationships are directed and use named connections between two named entities or nodes. For example, as shown in the preceding diagram, HAS_CEO is a relationship between **Company** and **Employee**. Relationships always have a direction, a type, a source node, and a target node. Relationships can also have properties. In the preceding example, the HAS_CEO relationship has a property called start_date.

Just like SQL standards, which are used to query RDBMS, graph databases can be queried using GQL. GQL is a newly announced ISO standard that helps query graph databases. One of the more popular open source GQLs available is openCypher. (You can learn more about openCypher at `https://opencypher.org/`.) Other popular graph database query languages include Cypher, TinkerPop3, and SPARQL.

Some examples of graph databases are Neo4J, ArangoDB, RedisGraph, Amazon Neptune, and GraphDB.

When to use:

- Fraud call detection.

- Recommendation engines.

- Customer engagement on travel websites.

- Referral relationships. For example, using a graph database, a healthcare provider can identify the various other providers they can get a referral from. This helps target specific clients and build a relationship that can be beneficial for both providers.

- Helps in a marketing campaign to identify influencers in a connected network by querying the number of incoming connections to a particular node.

In this section, we discussed various types of databases and when they should be used. We covered a few examples and sample use cases where a particular database should be chosen and why. In the next section, we will look at a few considerations a data architect should keep in mind while designing data models for various databases.

Data model design considerations

In this section, we will briefly discuss various design considerations you should consider while designing a data model for the various databases discussed in the previous section. The following aspects need to be considered while designing a data model:

- **Normalized versus denormalized**: Normalization is a data organization technique. It is used to reduce redundancy in a relationship or set of relationships. This is highly used in RDBMS, and it is always a best practice in RDBMS to create a normalized data model. In a normalized data model, you store a column in one of the tables (which is most suitable), rather than storing the same column in multiple tables. When fetching data, if you need the data of that column, you can join the tables to fetch that column. The following diagram shows an example of normalized data modeling using the crows-feet notation:

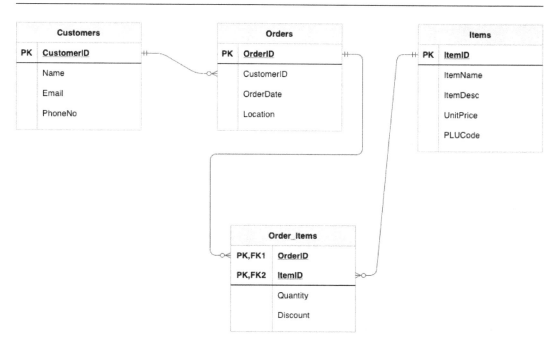

Figure 2.11 – Normalized data modeling

In the preceding diagram, none of the columns are repeated or redundant. Now, suppose we need to show an order that should display customer name, Customer ID, Order ID, item name, and Order date. To fetch this information, we can write a join query, like this:

```
SELECT cust.CustomerID, orders.OrderID, items.
ItemName,orders.OrderDate FROM Orders orders JOIN
Customers cust ON orders.CustomerID=cust.CustomerID JOIN
Order_items orderitem on orderitem.OrderID = orders.
OrderID JOIN Items items ON items.ItemID = orderitem.
ItemID
```

On the other hand, if we are designing the same thing for a NoSQL database, the focus should not be on reducing redundancy or normalizing the data. Instead, we should focus on the read speed. Such a change in design mindset is triggered by two important factors. First, NoSQL works with a huge volume of data and stores the data in distributed commodity hardware. So, data storage is not costly, and joins may not work efficiently if the volume of data is in hundreds of TBs or PBs. Second, NoSQL doesn't have the JOIN kind of queries, because NoSQL databases are non-relational. The following is an example of a document data model storing the same information that needs to be fetched:

```
{
    OrderId: Int, //documentKey
```

```
CustomerID: Int,
OrderDes: String
Items: [{
  itemId: Int,
  itemName: String,
  OrderDate: Date
  }]
}
```

As we can see, a single document contains all the necessary information, so this means a lot of redundant data. However, for big data scenarios, NoSQL works perfectly fine and provides great performance.

- **Query-first versus schema-first pattern**: While designing a data model in NoSQL, you must ask yourself what queries are going to run on this data model. The data model design in NoSQL usually starts with the kind of analytical query that will run on the model. This helps design a correct key for a document or a record in a NoSQL database. Also, in the case of a columnar database, it helps group columns in a column family based on the query that will run on the data. Such optimization helps NoSQL databases run queries with awesome performance on big data or unstructured data.

  On the other hand, RDBMS is designed to store data in a predefined schema that has been normalized and the relationships are very well defined. Since SQL is a declarative language and can query tables along with any related table during runtime, queries are not considered while designing an RDBMS data model.

- **Cost versus speed optimization**: With the advent of cloud databases and cloud-based solutions, understanding the cost considerations is a very important factor for a modern data architect. For example, when it comes to storage versus **Input/Output Operations Per Second** (**IOPS**), IOPS are always costlier than storage in cloud-based models. However, understanding the difference in how IOPS is calculated for an RDBMS or document store can help you save costs and effort in the longer term. An RDBMS IOPS is based on the page or the block size. So, RDBMS IOPS is determined by the number of pages it has accessed. However, in a document database, IOPS is based on the number of DB read/writes that happen in that database.

  Another example is that if, in an AWS DocumentDB, you give a greater number of indexes, you might get a better speed, but too many indexes will increase IOPS considerably, so it might cost you more. A safe limit of indexes per collection is five.

- **Indexes**: If you have a database, where you have a huge number of reads and you need to have great read performance, then you should consider having indexes in your database. Indexes help improve your read and update performance in your database. On the other hand, if you have a write-heavy application, indexes can slow down your insert performance.

- **Data distributions**: NoSQL databases are based on the scale-out architecture, where the data is stored and distributed across commodity nodes. One of the reasons that NoSQL databases have great performance for huge data volumes is that they can read or write data in parallel in the distributed nodes. However, if not designed properly, the data can be stored unevenly, which can cause a huge volume of data to be present in one node. This kind of uneven distribution of data in a distributed database is called **data skew**.

  Often, the problem of a node containing unusually high amounts of data, which can cause read or write bottlenecks for the database, is called **hotspotting**. Often, this happens due to a lack of understanding of the design principles of NoSQL databases and poor key design. In columnar databases, choosing an incremental sequence number as a key often leads to hotspotting. Instead, in both document and columnar databases, unique keys should be chosen and a combination of a few key column values should be concatenated in a particular order, preferably at least one of them being a text value. Techniques such as salting and MD5 encryption are used while designing the keys to help avoid hotspotting.

In this section, we covered the most obvious design considerations you should look at after you have chosen a database. While these considerations are basic for any data model design, there are other finer data model design techniques that are specific to the database you are choosing. We strongly recommend that you go over the official documentation of the database you've chosen for your solutions before you design your data model.

Summary

In this chapter, we covered the various data types and data formats that are available. We also discussed the various popular data formats that are used in modern data engineering and the compression techniques that are compatible with each. Once we understood the data types and formats, we explored various data storage formats – file, block, and object storage – we can use to store the data. Then, we discussed various kinds of enterprise data repositories in detail – data lake, data warehouse, and data marts. Once we covered the basics of data, including the different types and their storage, we briefly discussed databases and their types. We discussed various examples of databases, the USP of each kind of database, and when a particular kind of database should be chosen over another. We explored possible use cases when a database should be used.

Finally, we briefly covered the basic design considerations that a data architect should keep in mind while designing their data model using any chosen database.

Now that you know about data types, formats, databases, and when to use what, in the next chapter, we will explore the various platforms where data engineering solutions can be deployed and run.

3
Identifying the Right Data Platform

In the previous chapter, we discussed the various data types, their formats, and their storage. We also covered different databases and provided an overview of them. Then, we understood the factors and features we should compare when choosing a data format, storage type, or database for any use case to solve a data engineering problem effectively.

In this chapter, we will look at the various kinds of popular platforms that are available to run data engineering solutions. You will also learn about the considerations you should make as an architect to choose one of them. To do so, we will discuss the finer details of each platform and the alternatives these platforms provide. Finally, you will learn how to make the most of these platforms to architect an efficient, robust, and cost-effective solution for a business problem.

In this chapter, we're going to cover the following main topics:

- Virtualization and containerization platforms
- Hadoop platforms
- Cloud platforms
- Choosing the correct platform

Technical requirements

To complete this chapter, you'll need the following:

- JDK 1.8 or above
- Apache Maven 3.3 or above

The code for this chapter can be found in this book's GitHub repository: `https://github.com/PacktPublishing/Scalable-Data-Architecture-with-Java/tree/main/Chapter03`

Virtualization and containerization platforms

With the spread of **information technology** (IT) in all spheres of life, the dependency and reliability on IT infrastructure have increased manifold. Now, IT runs so many critical and real-time businesses. This means that there can be zero or negligible downtime for maintenance or failure. Also, rapid real-time demands have grown. For example, during the holiday season, there's a huge amount of traffic on online shopping websites. So, now, IT needs to be highly available, elastic, flexible, and quick. These were the reasons that motivated the creation of virtual platforms such as virtualization and containerization. For example, Barclays, a multinational financial firm based in the UK, was facing a hard time from competitors due to their slow pace of innovation and project deliveries. One of its major roadblocks was the time it took to provision new servers. So, they decided to use Red Hat OpenShift to containerize their application. This reduced the provisioning time dramatically from weeks to hours. As a result, time to market became super fast, which helped Barclays stay ahead of its competitors.

Virtualization abstracts the hardware and allows you to run multiple operating systems on a single server or piece of hardware. It uses software to create a virtual abstraction over the hardware resources so that multiple **virtual machines** (VMs) can run over the physical hardware with their *virtual OS*, **virtual CPU** (**vCPU**), virtual storage, and virtual networking. The following diagram shows how virtualization works:

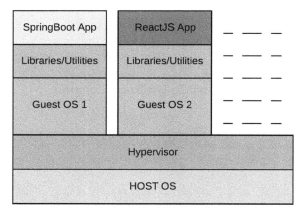

Figure 3.1 – Virtualization

As shown in the preceding diagram, VMs run on a host machine with the help of a hypervisor. A hypervisor is a piece of software or firmware that can host a VM on physical hardware such as a server or a computer. The physical machine where hypervisors create VMs are called host machines and the VMs are called guest machines. The operating system in the host machine is called the host OS, while the operating system in the VMs is called the guest OS.

Benefits of virtualization

The following are the benefits of virtualization:

- **Better resource utilization**: As multiple VMs run on the same hardware, hardware resources such as storage/memory and network can be more efficiently used to serve more applications that might have high loads at different times. Since spawning a VM is much quicker than spawning a new server. VMs can be spawned during high demand load cycles and switched off when the load on the application comes down.

- **Less downtime/higher availability**: When physical servers have issues or go down, need routine maintenance, or require upgrades, it results in costly downtime. With virtual servers, applications can readily move between guest hosts to make sure there is minimal downtime in the order of minutes rather than hours or days.

- **Quicker to market and scalability**: Since provisioning a VM takes minutes rather than weeks or months, the overall software delivery cycles have reduced substantially. This enables quicker testing since you can mock up production environment using VMs.

- **Faster disaster recovery (DR)**: Unlike physical servers, whose DR takes hours or days, VMs can recover within minutes. Hence, VMs enable us to have faster DR.

The following are a few examples of popular VMs:

- Microsoft's Hyper-V
- VMware's vSphere
- Oracle's VirtualBox

Let's see how a VM works. We will start this exercise by downloading and installing Oracle VirtualBox:

1. Based on your host operating system, you can download the appropriate installer of Oracle VirtualBox from `https://www.virtualbox.org/`.

   Then, install VirtualBox using the installation instructions at `https://www.virtualbox.org/manual/ch02.html`. These instructions are likely to vary by OS.

2. Once it has been installed, open Oracle VirtualBox. You will see the **Oracle VM VirtualBox Manager** home page, as shown here:

Figure 3.2 – The Oracle VirtualBox Manager home page

3. Then, click the **New** button to create a new VM on your machine (here, this serves as a guest OS). The following screenshot shows the **Create Virtual Machine** dialog screen popup (which appears upon clicking the **New** button):

Figure 3.3 – Configuring the guest VM using Oracle VirtualBox

In the preceding screenshot, you can see that you need to provide a unique name for the VM. You can also select the OS type and its version, as well as configure the memory (RAM) size. Finally, you can choose to configure or add a new virtual hard disk. If you choose to add a new hard disk, then a popup similar to the following will appear:

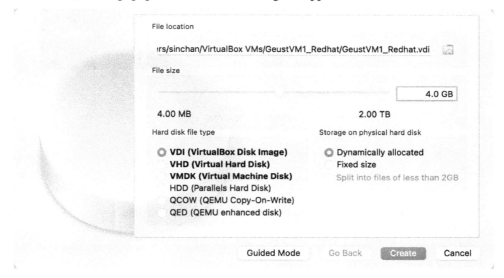

Figure 3.4 – Creating a virtual hard disk using Oracle VirtualBox

As shown in the preceding screenshot, when configuring a virtual hard disk, you can choose from various kinds of available virtual hard disk drives. The major popular virtual hard disks are as follows:

- **VirtualBox Disk Image (VDI)**
- **Virtual Hard Disk (VHD)**
- **Virtual Machine Disk (VMDK)**

Once you have configured your desired virtual hard disk configuration, you can create the VM by clicking the **Create** button.

4. Once a VM has been created, it will be listed on the **Oracle VM VirtualBox Manager** screen, as shown in the following screenshot. You can start the virtual machine by selecting the appropriate VM and clicking the **Start** button:

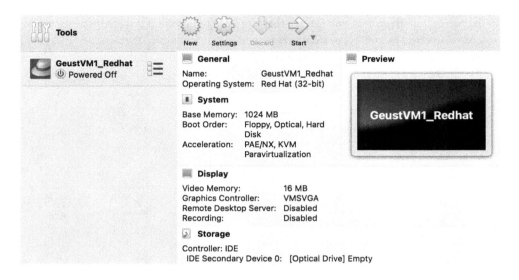

Figure 3.5 – Guest VM created and listed in Oracle VirtualBox

Although VMs simplify our delivery and make the platform more available and quicker than traditional servers, they have some limitations:

- **VMs are heavyweight components**: This means that every time you are doing a disaster recovery, you need to acquire all the resources and boot the guest operating system so that you can run your application. It takes a few minutes to reboot. Also, it is resource-heavy to boot a new guest OS.

- **They slow down the performance of the OS**: Since there are only a few resources in VMs, and they can only be thick provisioned, this slows down the performance of the host OS, which, in turn, affects the performance of the guest OS.

- **Limited portability**: Since the applications running on the VMs are tightly coupled to the guest OS, there will always be a portability issue when moving to a different guest OS with a different type or configuration.

Containerization can help us overcome these shortcomings. We'll take a look at containerization in the next section.

Containerization

Containerization is a technique that abstracts the OS (instead of the hardware) and lets applications run on top of it directly. Containerization is more efficient than virtualization as applications don't need a guest OS to run. Applications use the same kernel of the host OS to run multiple applications targeted for different types of OS. The following diagram shows how containerization works:

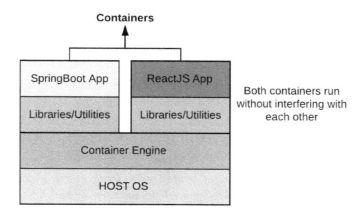

Figure 3.6 – Containerization

In containerization, a piece of software called a **container engine** runs on the host OS. This allows applications to run on top of the container engine, without any need to create a separate guest OS. Each running instance of the application, along with its dependencies, is called a container. Here, the application, along with its dependencies, can be bundled into a portable package called an image.

Benefits of containerization

The following are the advantages of containerization over virtualization:

- **Lightweight**: Containers use dependencies and binaries to run applications directly on a container engine. Containers don't need to create VMs, so they don't need to initialize a dedicated virtual memory/hard disk to run the application. Containers boot significantly faster than VMs. While VMs take minutes to boot, containers can boot in seconds.

- **Portable**: Applications, along with their dependencies and base container, can be bundled in a package called an image that can easily be ported across any container engine run on any kind of host.

- **Reduces single points of failure**: Due to the easy portability and lightweight nature of containers, testing, deploying, and scaling applications has become easier. This has led to the development of microservices, which ensures reduced single points of failure.

- **Increased development velocity**: In containerization, containers can be seamlessly migrated from one environment to another, enabling seamless continuous deployments. It also enables on-the-fly testing while building and packaging, hence improving continuous integration workflows. Application scaling becomes super fast if the application is run on containers. These features have made development easier and faster, enabling businesses to deliver solutions to the market quickly.

Docker is the most popular container engine. Let's look at some of the most important and common terms related to Docker:

- **Docker image**: A Docker image is a blueprint or template with instructions to create a Docker container. We can create a Docker image by bundling an already existing image with an application and its dependencies.

- **Docker container**: A Docker container is a running instance of a Docker image. A Docker container contains a write layer on top of one or more read layers. The writable layer allows us to write anything on the container, as well as execute commands.

- **Docker registry**: This is a repository that stores Docker images developed and uploaded by developers to be leveraged by other developers. Container repositories are physical locations where your Docker images are stored. Related images with the same name can be also stored, but each image will be uniquely identified by a tag. Maven repositories are to Java artifacts what Docker registries are to Docker images. Just like a Maven repository supports multiple versions of the related JAR files with the same name, a Docker registry supports multiple tags for images with the same name. Docker Hub is the official cloud-based public Docker registry.

- **Docker networking**: Docker networking is responsible for communication between the Docker host and Docker applications. It is also responsible for basic inter-container communication. External applications and developers can access applications running in a Docker container on the port exposed to the external world.

- **Docker storage**: Docker has multiple storage drivers that allow you to work with the underlying storage devices, such as Device Mapper, AUFS, and Overlay. Data volumes can be shared across multiple Docker containers. This enables shared resources to be stored in such data volumes. However, there is no way to share memory between Docker containers.

Now that we have learned about the important terminologies related to Docker, let's learn how to set up Docker in a local machine using Docker Desktop. We will also show you how to deploy an image and start a Docker container:

1. First, you must install Docker Desktop on our local machine. You can download the appropriate Docker Desktop version from the following link based on your OS type and version: `https://www.docker.com/products/docker-desktop`.

   Based on your operating system, you can follow the installation instructions at `https://docs.docker.com/desktop/mac/install/` (for Mac) or `https://docs.docker.com/desktop/windows/install/` (for Windows).

   If you don't have Maven installed in your system, please download and install it (instructions for installing Maven can be found at `https://maven.apache.org/install.html`).

2. Once you have installed Docker Desktop, open it. You will be asked to accept an agreement. Please read and accept it. Once you have done this and the application opens, you will see the following home page:

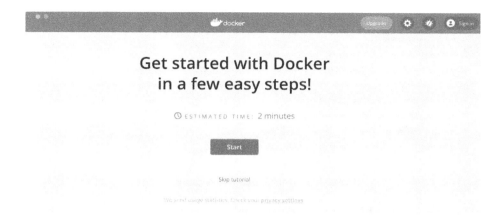

Figure 3.7 – Docker Desktop home page

3. Next, you must create a Docker Hub personal account to make use of Docker Desktop efficiently. Please sign up to create a personal Docker account at https://hub.docker.com/signup.

    Once you have successfully created your account, click the **Sign In** button and enter your Docker ID and password to log in, as shown in the following screenshot:

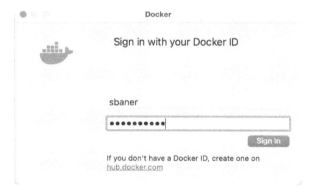

Figure 3.8 – Logging into Docker Desktop

4. Now, let's build our own Docker file. To build a Docker file, we need to know basic Docker build commands. The following table lists a few important Docker build commands:

Command	Description
FROM	To specify the base image.
EXPOSE	To define which port will be used to access your container application.
ENTRYPOINT	A command that will always be executed when the container starts. If not specified, the default is /bin/sh -c
CMD	Arguments passed to the entrypoint. If ENTRYPOINT is not set (defaults to /bin/sh -c), the CMD will be the command the container executes.
COPY	To copy over files or directories from a specific location.
ADD	The same as COPY, but also able to handle remote URLs and unpack compressed files.
RUN	To install any applications and packages required for your container.
WORKDIR	To set the working directory for any commands that follow in the Dockerfile.

Figure 3.9 – Docker build commands

To build the Docker file, first, download the code from `https://github.com/PacktPublishing/Scalable-Data-Architecture-with-Java/blob/main/Chapter03/sourcecode/DockerExample`. In this project, we will create a simple REST API using Spring Boot and deploy and run this application using our Docker environment locally. The artifact that will be generated when we build this project is `DockerExample-1.0-SNAPSHOT.jar`. The Docker file will look as follows:

```
# Each step creates a read-only layer of the image.
# For Java 8
FROM openjdk:8-jdk-alpine

# cd /opt/app
WORKDIR /opt/app

# cp target/DockerExample-1.0-SNAPSHOT.jar /opt/app/app.
jar
COPY target/DockerExample-1.0-SNAPSHOT.jar app.jar

# exposing the port on which application runs
EXPOSE 8080

# java -jar /opt/app/app.jar
ENTRYPOINT ["java","-jar","/app.jar"]
```

In *step 1* of the Docker file's source code, we import a base image from Docker Hub. In *step 2*, we set the working directory inside the Docker container as /opt/app. In the next step, we copy our artifact to the working directory on Docker. After that, we expose port 8080 from Docker. Finally, we execute the Java app using the java -jar command.

5. Next, we will build the JAR file using Maven. First, using the command line (Windows) or a Terminal (Mac), go to the root folder of the `DockerExample` project. Run the following command to build the JAR file from the root folder of the project:

```
> mvn clean install
```

6. Then, run the following command to create a customized Docker image named `hello-docker` from the Docker file we just created:

```
> docker build --tag=hello-docker:latest .
```

Once you run this command, you will be able to see the Docker image in Docker Desktop, in the **Images** tab, as shown in the following screenshot:

Figure 3.10 – Docker container created successfully and listed

7. Now, you can start the container by clicking the **RUN** button, as follows:

Figure 3.11 – Running a Docker container

Provide a container name and a host port value and click **Run** in the popup dialog to start the container, as shown in the following screenshot:

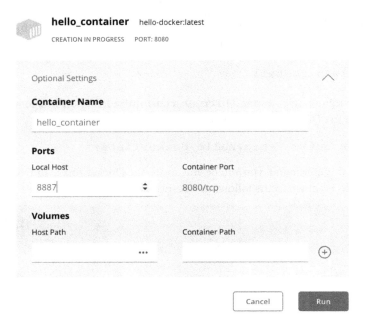

Figure 3.12 – Setting up Docker Run configurations

Once you click **Run**, the container will be instantiated and you will be able to see the container listed (with its status set to **RUNNING**) in the **Containers / Apps** tab of Docker Desktop, as shown in the following screenshot:

Figure 3.13 – Running instance on Docker Desktop

8. Now, you can validate the app by testing it in a browser. Please make sure you are using the host port (configured during container creation) in the HTTP address while validating the application. For our example, you can validate the application using port 8887 (which you mapped earlier), as follows:

Figure 3.14 – Testing the app deployed on Docker

You can log into the Docker CLI using the **CLI** button in Docker Desktop, as follows:

Figure 3.15 – Opening the Docker CLI from Docker Desktop

In this section, we learned about Docker. While Docker makes our lives easy and makes development and deployment considerably faster, it comes with the following set of challenges:

- Inter-container communication is usually not possible or very complex to set up
- There is no incoming traffic distribution mechanism, which might cause a skewed distribution of incoming traffic to a set of containers
- Container management is overhead to manage the cluster manually
- Auto-scaling is not possible

In a production environment, we need to solve these shortcomings if we want to run a robust, efficient, scalable, and cost-effective solution. Here, container orchestrators come to the rescue. There are many container orchestrators on the market. However, Kubernetes, which was developed and open-sourced by Google, is one of the most popular and widely used container orchestrators. In the next section, we will discuss Kubernetes in more detail.

Kubernetes

Kubernetes is an open source container orchestrator that effectively manages containerized applications and their inter-container communication. It also automates how containers are deployed and scaled. Each Kubernetes cluster has multiple components:

- Master
- Nodes
- Kubernetes objects (namespaces, pods, containers, volumes, deployments, and services)

The following diagram shows the various components of a Kubernetes cluster:

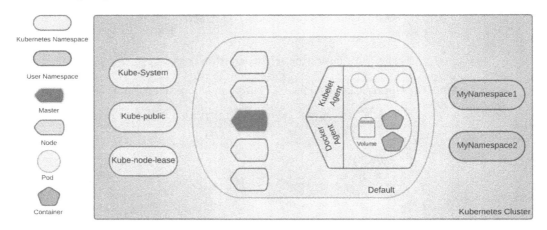

Figure 3.16 – A Kubernetes cluster and its components

Now, let's briefly describe each of the components shown in the preceding diagram:

- **Kubernetes namespace**: This provides logical segregation between different applications for different teams that are built for different purposes. **Kube-System**, **Kube-public**, and **Kube-node-release** are namespaces that are used by Kubernetes internal systems to manage the cluster, such as reading heartbeats, and to store publicly accessible data, such as configMap. Apart from this, there is a default namespace.

- **User namespace**: Users/teams can create a namespace inside the default namespace.

- **Master nodes**: The master node is the cluster orchestrator. It scales and allocates app containers whenever a new request for a deployment comes into the Kubernetes cluster.

- **Worker nodes**: These are the nodes where the pods are deployed and the applications run.

- **Pod**: A pod is an abstraction on top of a container that helps it become easily portable between different runtimes, auto-detects the available ports in a cluster, and gives it a unique IP address. A pod can consist of multiple containers in which multiple helper containers can communicate seamlessly and assist the primary application. These multiple containers in a single pod not only share volumes but also share memory spaces such as **Portable Operating System Interface** (**POSIX**) shared memory.

- **Agents**: There are two types of agents, as follows:

- **Kubelet agent**: This is a service that runs in each node. It ensures all the containers within that node are up and running.

- **Docker agent**: This is a service that is used to run a container.

Now that we have briefly seen the components of Kubernetes and its role in containerization, let's try to deploy the Docker image we created in the previous section in a Kubernetes cluster locally. To do that, we must install **minikube** (a Kubernetes cluster for running Kubernetes on your local machine):

1. You can install the appropriate version of `minikube` by following the instructions at `https://minikube.sigs.k8s.io/docs/start/`.

2. Once `minikube` has been installed, you can start `minikube` using the following command:

    ```
    > minikube start
    ```

A successful `minikube start` looks as follows:

```
sinchan@Sinchans-MBP-2 Downloads % minikube start
   minikube v1.23.2 on Darwin 11.2 (arm64)
   Automatically selected the docker driver. Other choices: virtualbox, ssh
   Starting control plane node minikube in cluster minikube
   Pulling base image ...
   Downloading Kubernetes v1.22.2 preload ...
   > preloaded-images-k8s-v13-v1...: 541.26 MiB / 541.26 MiB  100.00% 18.98 Mi
   > gcr.io/k8s-minikube/kicbase: 321.23 MiB / 321.23 MiB  100.00% 8.45 MiB p/
   Creating docker container (CPUs=2, Memory=1988MB) ...
   Preparing Kubernetes v1.22.2 on Docker 20.10.8 ...
    ■ Generating certificates and keys ...
    ■ Booting up control plane ...
    ■ Configuring RBAC rules ...
   Verifying Kubernetes components...
    ■ Using image gcr.io/k8s-minikube/storage-provisioner:v5
   Enabled addons: storage-provisioner, default-storageclass
   Done! kubectl is now configured to use "minikube" cluster and "default" namespace by default
```

Figure 3.17 – Starting minikube

3. Now, just like we have to create a Dockerfile to create a Docker image, we have to create a YAML file to give deployment instructions to the Kubernetes cluster. In our project, we will name this YAML file `deployment.yaml`. The following is the code in the `deployment.yaml` file:

    ```
    apiVersion: v1
    kind: Service
    metadata:
      name: hello-docker-service
    spec:
      selector:
        app: hello-docker-app
      ports:
        - protocol: "TCP"
    ```

```
        port: 8080
        targetPort: 8080
        nodePort: 30036
    type: LoadBalancer

---
apiVersion: apps/v1
kind: Deployment
metadata:
  name: hello-docker-app
spec:
  selector:
    matchLabels:
      app: hello-docker-app
  replicas: 5
  template:
    metadata:
      labels:
        app: hello-docker-app
    spec:
      containers:
        - name: hello-docker-app
          image: hello-docker-app
          imagePullPolicy: Never
          ports:
            - containerPort: 8080
```

The `deployment.yaml` file contains two types of configuration: one for `Service` and one for `Deployment`. Each Kubernetes component configuration consists of mainly three parts:

- `metadata` consists of the name and any other meta information.

- `spec` contains the specification. This is directly dependent on the kind of component that is being configured.

- `status` is not something we have to configure. The Kubernetes cluster adds that part and keeps updating it after the deployment is done.

4. First, you have to build the `hello-docker-app` image and expose it to the `minikube` Docker environment. You can do that by executing the following commands:

```
> eval $(minikube docker-env)
> docker build --tag hello-docker-app:latest .
```

5. Now, you can deploy this application in a Kubernetes cluster using the following command from the project root folder:

```
> kubectl apply -f deployment.yaml
```

After executing this command, you should be able to see that `hello-docker-service` and `hello-docker-app` were created successfully, as shown in the following screenshot:

```
sinchan@Sinchans-MacBook-Pro-2 DockerExample % kubectl get services
NAME                    TYPE           CLUSTER-IP      EXTERNAL-IP   PORT(S)          AGE
hello-docker-service    LoadBalancer   10.99.213.136   <pending>     8080:30036/TCP   42m
kubernetes              ClusterIP      10.96.0.1       <none>        443/TCP          178m
sinchan@Sinchans-MacBook-Pro-2 DockerExample % kubectl get deployments
NAME                READY   UP-TO-DATE   AVAILABLE   AGE
hello-docker-app    5/5     5            5           42m
```

Figure 3.18 – Applications created in Kubernetes cluster

6. You can also check the deployments and their status in the minikube dashboard by executing the following command:

```
> minikube dashboard
```

Once executed, you should see the dashboard appear in your default browser. Here, you will be able to see the status of your deployment, as well as other monitoring information:

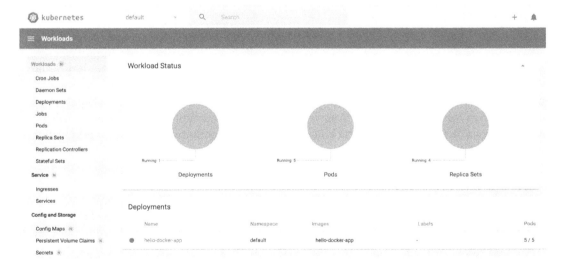

Figure 3.19 – The minikube dashboard

7. Now, you can access the deployed application and start its services using the following command:

    ```
    > minikube start service: hello-docker-service
    ```

    Once you have started the service, you can check the base URL that's been exposed by the Docker service linked to your application by executing the following command:

    ```
    > minikube service --url hello-docker-service
    ```

    This command will show an output similar to the following:

```
sinchan@Sinchans-MacBook-Pro-2 DockerExample % minikube service --url hello-docker-service
🏃  Starting tunnel for service hello-docker-service.
|-----------|----------------------|--------------|-------------------------|
| NAMESPACE |         NAME         | TARGET PORT  |           URL           |
|-----------|----------------------|--------------|-------------------------|
|  default  | hello-docker-service |              | http://127.0.0.1:63883  |
|-----------|----------------------|--------------|-------------------------|
http://127.0.0.1:63883
```

Figure 3.20 – Checking the base URL

8. You can verify the running application from your browser by using the `http://127.0.0.1:63883/hello` URL for this example, as follows:

Figure 3.21 – Testing app deployed using Kubernetes

In this section, we discussed how virtualization and containerization can help you manage, deploy, and develop an application in more effective, faster, and cost-optimized ways. General web applications, backend applications, and other processing applications work extremely well on scalable virtual platforms such as containers and VMs. However, big data, which amounts to terabytes and petabytes of data, requires a platform with a different kind of architecture to perform well. From the next section onwards, we will discuss platforms that are apt for big data processing.

Hadoop platforms

With the advent of search engines, social networks, and online marketplaces, data volumes grew exponentially. Searching and processing such data volumes needed a different approach to meet the **service-level agreements (SLAs)** and customer expectations. Both Google and Nutch used a new technology paradigm to solve this problem, thus storing and processing data in a distributed way automatically. As a result of this approach, Hadoop was born in 2008 and has proved to be a lifesaver for storing and processing huge volumes (in the order of terabytes or more) of data efficiently and quickly.

Apache Hadoop is an open source framework that enables distributed storage and processing of large datasets across a cluster of computers. It is designed to scale from a single server to thousands of machines easily. It provides high availability by having strong node failover and recovery features, which enables a Hadoop cluster to run on cheap commodity hardware.

Hadoop architecture

In this section, we will discuss the architecture and various components of a Hadoop cluster. The following diagram provides a top-level overview of the Hadoop ecosystem:

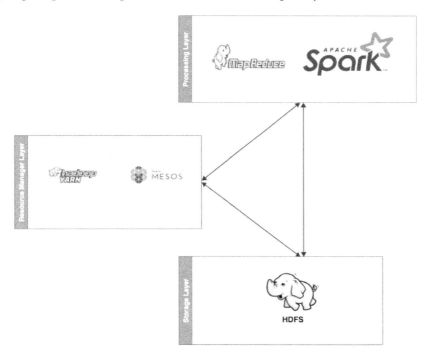

Figure 3.22 – Hadoop ecosystem overview

As shown in the preceding diagram, the Hadoop ecosystem consists of three separate layers, as discussed here:

- **Storage layer**: The storage layer in Hadoop is known as **Hadoop Distributed File System (HDFS)**. HDFS supports distributed and replicated storage of large datasets, which provides high availability and high-performance access to data. The following diagram provides an overview of the HDFS architecture:

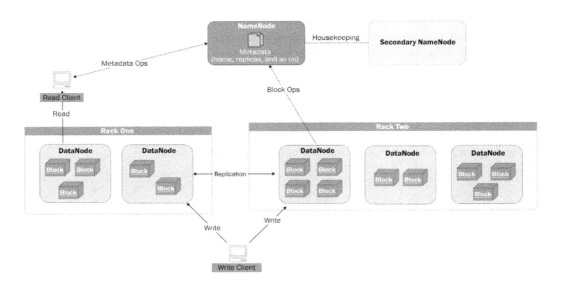

Figure 3.23 – HDFS architecture

HDFS has a master-slave architecture where the **NameNode** is the master and all **DataNodes** are the slaves. The **NameNode** is responsible for storing metadata about all the files and directories in HDFS. It is also responsible for storing a mapping of which blocks are stored in which **DataNode**. There is a secondary **NameNode** that is responsible for the housekeeping jobs of the **NameNode** such as compaction. DataNodes are the real horsepower in an HDFS system. They are responsible for storing block-level data and performing all the necessary block-level operations on it. The **DataNode** sends periodical signals called heartbeats to the **NameNode** to specify that they are up and running. It also sends a block report to the **NameNode** every tenth heartbeat.

When the client makes a read request, it gets the metadata information about the files and blocks from the **NameNode**. Then, it fetches the required blocks from the correct **DataNode**(s) using this metadata. When a client makes a write call, the data gets written into distributed blocks across various DataNodes. These blocks are then replicated across the nodes (on a different rack) for high availability in case there is an outage in the current rack.

- **Resource manager layer**: The resource manager is a framework that manages cluster resources and is also responsible for scheduling Hadoop jobs. The following diagram shows how the resource manager works:

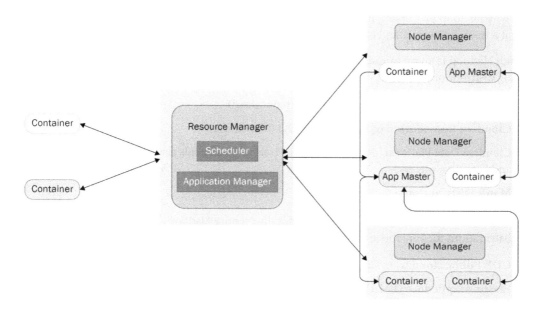

Figure 3.24 – How resource manager works

As we can see, each client sends a request to the **Resource Manager** when they submit a processing job in Hadoop. The **Resource Manager** consists of a **Scheduler** and an **Application Manager**. Here, the **Application Manager** is responsible for negotiating with the application's master container by communicating with node managers in different data nodes. Each application master is responsible for executing a single application. Then, the **Scheduler** in the **Resource Manager** is responsible for negotiating other container resources by interacting with the node managers based on the resource requests from the ApplicationMaster. Two of the most popular resource managers in Hadoop are **Apache YARN** and **Apache Mesos**.

- **Processing layer**: The processing layer is responsible for the parallel processing of distributed datasets in the Hadoop ecosystem. Two of the most popular Hadoop processing engines are **MapReduce** and **Apache Spark**. MapReduce programs are tightly coupled to the Hadoop environment. It primarily processes data using two mandatory phases – the map phase and the reduce phase – and uses several optional data processing phases. It writes intermediate data back to HDFS between these phases. On the other hand, Spark reads the distributed data in a logically distributed dataset called **Resilient Distributed Dataset** (RDD) and creates a **Directed Acyclic Graph** (DAG) consisting of stages and tasks to process the data. Since it doesn't write the intermediate data to disk (unless shuffling is explicitly required), it usually works 10 times faster than MapReduce.

Although these three layers are interdependent, the design is such that the layers are decoupled from each other. This decoupled layer architecture makes Hadoop more flexible, powerful, and extendable.

This is why Hadoop processing has improved and evolved, even though the sizes of datasets have grown at a tremendous rate and expected SLAs to process data have reduced over time.

Although Hadoop is an open source framework, all production Hadoop clusters run in one of the following distributions:

- **Hortonworks Data Platform** or **HDP** (discontinued)

- **Cloudera Distribution of Hadoop** or **CDH** (discontinued)

- **Cloudera Data Platform** (both HDP and CDH can migrate to this platform after the Hortonworks and Cloudera merger)

- **MapR Distributions**

Apart from these distributions, which are meant for on-premise Hadoop deployments, some popular cloud distributions for Hadoop are available:

- **CDP Public cloud** from Cloudera

- **Elastic MapReduce (EMR)** from **Amazon Web Services (AWS)**

- **HDInsight** from Microsoft Azure

- **Cloud Dataproc** from **Google Cloud Platform (GCP)**

In this section, we briefly discussed Hadoop distributions and how they work. We also covered the various Hadoop distributions that are available from various vendors for running Hadoop in a production environment.

As data keeps growing, there is a need to grow the on-premise infrastructure. Such infrastructure capacities need to be planned to support the maximum load. This creates either underutilization of resources or overutilization of resources if an unexpected load occurs. The answer to this problem is cloud computing. In the next section, we will discuss various cloud platforms and the benefit they bring to data engineering solutions.

Cloud platforms

Cloud computing involves delivering computing services such as storage, compute, networking, and intelligence over the internet. It offers a pay-as-you-go model, which means you only pay for the service you use. This helps cut down on your operating costs, as well as **capital expenditure (CapEx)** costs. Cloud enables optimal resource utilization, instant scalability, agility, and ease of maintenance, enabling faster innovation and economies of scale. For example, Canva is a design tool that anyone can access via its simple user interface. In 2019, it had 55 million users. At the time of writing, it has 85 million users worldwide, creating 100+ designs per second. To accommodate this exponential customer and data volume growth seamlessly with similar or better performance, Canva uses the AWS platform.

The following cloud computing distributions are the market leaders in cloud computing and are often referred to as the Big 3 of cloud computing:

- AWS by Amazon

- Microsoft Azure by Microsoft

- GCP by Google

Apart from the Big 3, there are other smaller or lesser-known cloud distributions such as Red Hat OpenShift, HPE GreenLake, and IBM Cloud.

Benefits of cloud computing

The following are the benefits of cloud computing:

- **Cost-effective**: Cloud computing reduces the CapEx cost by eliminating the huge costs involved in the infrastructure setup. It also reduces cost by applying a pay-as-you-go model.

- **Scalable**: Since cloud services are all virtualized, they can be spun up within minutes or seconds, enabling extremely fast scalability.

- **Elastic**: Cloud services can be easily scaled up or down based on the resource and compute demand.

- **Reliable**: Since each service is replicated across availability zones as well as regions, the services are highly reliable and guarantee minimum downtime.

- **Global**: Since the cloud service is on the internet, the computation power can be delivered across the globe, from the right geographic location.

- **Increased productivity**: Since provisioning, managing, and deploying resources and services are no longer headaches for developers, they can focus on business functionality and deliver solutions much faster and effectively.

- **Secure**: Along with the other benefits, the cloud has a lot of security layers and services, making the cloud a secure infrastructure.

There are three types of cloud computing, as follows:

- **Public cloud**: Public clouds are owned and operated by third-party vendors who deliver computing resources and services over the internet. Here, as a user of the public cloud, you must pay for what you use.

- **Private cloud**: A private cloud is a form of cloud computing where the computing resources are owned by the customer, usually in a private on-premise data center. The cloud provider only provides the cloud software and its support. Usually, this is used by big enterprises where security and compliance are constraints for moving toward the public cloud. Here, the customers are responsible for managing and monitoring cloud resources.

- **Hybrid cloud**: Hybrid clouds combine both public and private clouds, bound together by technology through which data and applications can communicate and move seamlessly between private and public clouds. This provides higher flexibility when it comes to security, compliance, and agility.

Now that we have discussed the different types of cloud computing, let's try to understand the various types of cloud services available in a public cloud distribution. The various types of cloud services are as follows:

- **Infrastructure as a Service (IaaS)**

- **Platform as a Service (PaaS)**

- **Software as a Service (SaaS)**

In the cloud, the responsibility of owning various stacks in application development is shared between cloud vendors and the customers. The following diagram shows the shared responsibility model for these kinds of cloud computing services:

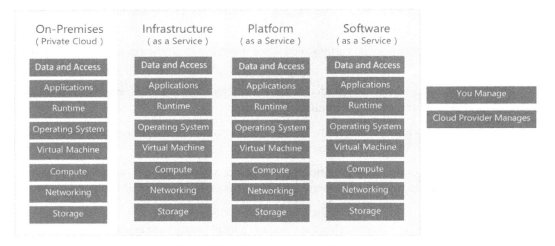

Figure 3.25 – Shared responsibility model

As we can see, if the customer is running a private cloud, all the resources, services, applications, and data are the customer's responsibility. However, if you opt for a public cloud, then you can choose between IaaS, PaaS, and SaaS. Cloud vendors promise to manage and own infrastructure services such as compute, storage, and networking in an IaaS model. If you go for a PaaS model, apart from what you get in IaaS, cloud providers also manage the OS, VMs, and runtime so that you can own, develop, and manage your applications, data, and access. In SaaS, everything except data and access is managed by your cloud vendor. Even the application or software is managed by the cloud provider. Although this might be costlier if you take a single unit compared to the other two models, based on your business, it might be cheaper and more hassle-free.

With that, we have discussed the various platforms where data engineering applications may be deployed. Now, let's discuss the various design choices that an architect needs to know to choose the correct platform for them.

Choosing the correct platform

In this section, we will look at one of the most important decisions architects have to make – *how to choose the most suitable platform for a use case*. Here, we will understand when to choose between virtualization versus containerization and on-premise versus the cloud when considering various cloud data platforms.

When to choose virtualization versus containerization

Although both these technologies ensure that we can use resources to the best of our ability by provisioning virtual resources, each has its advantages based on the type of application.

Microservices is a variant of the service-oriented architecture where an application is perceived as a collection of loosely coupled services. Each service is fine-grained and lightweight. Microservices are best suited for container-based platforms. For example, a REST service can be easily deployed using containers. Since microservices consist of loosely coupled services, they should be easily deployable and scalable. Since each service can be independently consumed and reused by other services and stacks, they need to be portable so that they can quickly migrate to any containerized platform.

On the other hand, monolithic applications are designed to perform multiple related tasks, but it is built as a tightly coupled single application. Such applications are more suited for small teams or **Proof of Concept** (**POC**) purposes. Another use case where such monolithic architectures are used is in legacy applications. Such monolithic applications are best suited for virtualization. Another use case where virtualized platforms are preferred over containerization is in any application that is dependent on an OS or talks directly to a specific OS.

However, in the cloud, all the servers that are provisioned are VMs. Containerized platforms such as Amazon **Elastic Container Service** (**ECS**) and Amazon **Elastic Kubernetes Service** (**EKS**) run on top of virtual servers such as Amazon **Elastic Compute Cloud** (**EC2**). So, in modern architectures, especially in the cloud, the question is not choosing between containerization and virtualization – it is choosing between containerization along with virtualization versus virtualization.

When to use big data

If we are handling data that is terabytes or petabytes in size, big data is a good choice. As **artificial intelligence** (**AI**) and **machine learning** (**ML**) applications are growing in popularity, we need to deal with huge volumes of data – the larger the data, more accurate will be the AI models. These volumes of data run into the terabytes. Processing such data in a scalable fashion can be done by big data applications. There are use cases where, due to processing complexity, processing hundreds of gigabytes of data takes an unnecessarily long time. In such scenarios, big data may be a good solution. Most big data use cases are for analytics, **Online Analytical Processing (OLAP)**, AI, and ML.

Choosing between on-premise versus cloud-based solutions

This is an obvious question that architects face today. In this section, we will try to see what factors affect this decision, as well as recommend a few general criteria to help you decide on one over the other. The factors that will help you decide between on-premise versus cloud solutions are as follows:

- **Cost**: Enterprises are responsible for all infrastructure and maintenance costs, including human resources for on-premise environments. Costs also include migration costs from on-premise to the cloud. Another important cost metric is about CapEx versus **operating expenses (OpEx)** and how cost-efficient OpEx will be versus CapEx for the enterprise. All these factors determine the total cost of ownership, which ultimately determines what is best for your business.

- **Control**: Enterprises are completely in control of the data, its storage, and all hardware related to the on-premise infrastructure. However, although enterprises own the data, the storage and its hardware are managed by the cloud provider.

- **Resource demand pattern**: If the demand for resources is elastic and infrastructure demand is seasonal, then the cloud may be the correct choice. On the other hand, if resource demand is static, then opting for on-premise may be the correct option.

- **Agility and scalability**: If your company is a start-up and growing exponentially, which means your demand scales up and down based on the feedback you receive and your volatile customer base, then the cloud will be a better choice for you.

- **Security**: Security is a big concern for a few industries, such as finance and healthcare. Although there have been many advances in cloud security and they have a strong robust security model, since the data is stored in hardware managed by a public cloud provider, many such businesses with very sensitive data choose on-premise over the cloud for security reasons.

- **Compliance**: Several industries have very strict regulatory controls and policies, such as federal agencies and healthcare. In such businesses, having complete control over the data and its storage makes more sense. Hence, on-premise options are more suitable.

Based on these factors, here are some broad guidelines that you can use to make this decision. However, note that these are only recommendations – the actual choice will depend on your specific business needs and context.

You should choose on-premise architectures in the following circumstances:

- Security is a major concern and you don't want to take the chance of any data risks occurring

- Regulatory policies and controls are stringent, stipulating that control of data and its storage should remain within the organization

- Legacy systems can't easily be moved or replicated
- The time, effort, and cost involved are not justifiable to migrate data and processing from on-premise to the cloud

You should choose cloud architectures in the following circumstances:

- Flexibility and agility to scale and grow are needed
- You are a start-up and you have a limited client base and limited CapEx, but you have high growth potential
- You want dynamic configurations of the environment that can easily be modified on demand
- You do not want to do a CapEx investment on infrastructure and prefer a pay-as-you-go model
- You are uncertain about the business demand, and you need to scale your resources up and down frequently
- You do not want to expend resources and time to maintain your infrastructure and the cost associated with it
- You want an agile setup, quicker deliveries, and a faster turnaround time for operations

Finally, let's compare the Big 3 cloud vendors to decide which provider is a best fit for your business.

Choosing between various cloud vendors

In this section, we will compare the Big 3 public cloud vendors and how they perform in various categories, even though there is no clear answer to the question, *Which cloud vendor is best for my business?* The following table provides a comparison between the Big 3 cloud providers and throws light on their strengths and weaknesses:

	AWS	Azure	GCP
Services	Huge range of services	Good range of services available. Exceptional services in AI/ML.	Limited services are available.
Maturity	Most mature	Catching up with AWS.	Still relatively less mature than the other two.
Marketplace	All vendors make their products available	Good vendor support but less than AWS.	
Reliability	Excellent	Excellent.	Excellent.
Security	Excellent	Excellent.	Fewer notches than AWS and Azure.
Cost	Varies	Most cost-efficient.	Varies.

	AWS	Azure	GCP
Support	Paid dev/enterprise support	Paid dev/enterprise support. More support options than AWS.	Paid dev/premium support. Costlier support than the other two.
Hybrid Cloud Support	Limited	Excellent.	Good.
Special Notes	More compute capacity versus Azure and GCP	Easy integration and migrations for existing Microsoft services.	Excellent support for containerized workloads. Global fiber network.

Figure 3.26 – Comparison of the Big 3 cloud vendors

In short, AWS is the market leader but both Azure and GCP are catching up. If you are looking for the maximum number of services available across the globe, AWS will be your obvious choice, but it comes with a higher learning curve.

If your use case revolves only around AI/ML and you have a Microsoft on-premise infrastructure, Azure may be the correct choice. They have excellent enterprise support and hybrid cloud support. If you need a robust hybrid cloud infrastructure, Microsoft Azure is your go-to option.

GCP entered the race late, but they have excellent integration and support for open source and third-party services.

But in the end, it boils down to your specific use case. As the market is growing, most enterprises are looking for multi-cloud strategies to leverage the best of each vendor.

Now, let's summarize this chapter.

Summary

In this chapter, we discussed various virtualization platforms. First, we briefly covered the architectures of the virtualization, containerization, and container orchestration frameworks. Then, we deployed VMs, Docker containers, and Kubernetes containers and ran an application on top of them. In doing so, we learned how to configure Dockerfiles and Kubernetes deployment scripts. After that, we discussed the Hadoop architecture and the various Hadoop distributions that are available on the market. Then, we briefly discussed cloud computing and its basic concepts. Finally, we covered the decisions that every data architect has to make: *containers or VMs? Do I need big data processing? Cloud or on-premise? If the cloud, which cloud?*

With that, we have a good understanding of some of the basic concepts and nuances of data architecting, including the basic concepts, databases, data storage, and the various platforms these solutions run on in production. In the next chapter, we will dive deeper into how to architect various data processing and data ingestion pipelines.

Section 2 – Building Data Processing Pipelines

This section focuses on guiding you to learn how to architect and develop batch processing and stream processing solutions using various technologies in the Java stack. Finally, it will also help you to understand and apply data governance and security to a solution practically.

This section comprises the following chapters:

4

ETL Data Load – A Batch-Based Solution to Ingesting Data in a Data Warehouse

In the previous chapters, we discussed various foundational concepts surrounding data engineering, starting with the different types of data engineering problems. Then, we discussed various data types, data formats, data storage, and databases. We also discussed the various platforms that are available to deploy and run data engineering solutions in production.

In this chapter, we will learn how to architect and design a batch-based solution for low to medium-volume data ingestion from a data source to a **data warehouse**. Here, we will be taking a real-time use case to discuss, model, and design a data warehouse for such a scenario. We will also learn how to develop this solution using a Java-based technical stack and run and test our solution. By the end of this chapter, you should be able to design and develop an **extract, transform, load** (ETL)-based batch pipeline using Java and its related stack.

In this chapter, we're going to cover the following main topics:

- Understanding the problem and source data
- Building an effective data model
- Designing the solution
- Implementing and unit testing the solution

Technical requirements

You can find all the code files for this chapter in this book's GitHub repository: `https://github.com/PacktPublishing/Scalable-Data-Architecture-with-Java/tree/main/Chapter04/SpringBatchApp/EtlDatawarehouse`.

Understanding the problem and source data

Data engineering often involves collecting, storing, and analyzing data. But nearly all data engineering landscapes start with ingesting raw data into a data lake or a data warehouse. In this chapter, we will be discussing one such typical use case and build an end-to-end solution for the problem discussed in the following section.

Problem statement

Company XYZ is a third-party vendor that provides services for building and maintaining data centers. Now, Company XYZ is planning to build a data center monitoring tool for its customer. The customer wants to see various useful metrics, such as the number of incidents reported for any device on an hourly, monthly, or quarterly basis. They also want reports on closure ratios and average closure duration. They are also interested in searching incidents based on the type of device or incident type. They are also interested to find time-based outage patterns to predict seasonal or hourly usage surges for any set of resources. These reports need to be generated once every 12 hours. To generate such reports, a data warehouse needs to be built, and data must be ingested and stored daily in that data warehouse so that such reports can easily be generated.

To create the solution for this data engineering problem, we must analyze the four dimensions of data (refer to the *Dimensions of data* section in *Chapter 1, Basics of Modern Data Architecture*) in this use case. Our first question would be, *What is the velocity of data?* The answer to this question helps us to determine whether it is a real-time or batch processing problem. Although there is not much information about the input frequency of data as per the problem statement, it is clearly stated that the report needs to be generated after every 12 hours or twice daily. Irrespective of the incoming speed of data, if the frequency in which the downstream system needs data is more than an hour, we can safely decide that we are dealing with a batch-based data engineering problem (please refer to the *Types of data engineering problems* section in *Chapter 1, Basics of Modern Data Architecture*).

Our second question would be, *What is the volume of the data? Is it huge? Is there a chance that this can grow into hundreds of terabytes in the future?* These questions generally help us choose the technology that we should use. If the volume is huge (in terabytes or hundreds of terabytes), only then should we choose **big data** technologies to solve our problem. A lot of times, architects tend to use big data in a non-big data use case, which makes the solution unsustainable and expensive in terms of cost, maintenance, and time. In our case, the data that needs to be ingested is incident log data. Such data is usually not huge. However, an architect should get confirmation about the data that will be sent for ingestion. In this case, let's suppose that the customers responded and said that the data will be sent every couple of hours as a flat file, consisting of a Delta of the incidents that have either been newly logged or updated. This would mean that our datasets will be either in a small file or a medium-sized file. This means that as an architect, we should choose a non-big data-based solution.

Understanding the source data

The third important question that an architect must ask is, *What is the variety of the data? Is it structured, unstructured, or semi-structured?* This question often helps to determine how such data can be processed and stored. If the data is unstructured, then we need to store it in a NoSQL database, but structured data can be stored in RDBMS databases. There is another question related to **data variety** – that is, *What is the format of the data? Is it in CSV format, JSON format, Avro format, or Parquet format? Is the data compressed when received?* Often, these questions help determine the techniques, technologies, processing rules, and pipeline design required to process and ingest the data. In our case, since it is not mentioned in the initial requirement, we need to ask the customers these questions. Let's suppose our customers agree to send the data in the form of CSV files. So, in this case, we are dealing with structured data and the data is coming as a CSV file without any compression. As it is structured data, it is apt for us to use a relational data model or RDBMS database to store our data.

This brings us to the final question regarding the dimension of the data: *What is the veracity of the data?* Or, in simpler terms, *What is the quality of the data that we receive? Is there too much noise in the data?* Of all the data engineering solutions that fail to solve a customer problem, the majority fail because of a lack of time spent analyzing and profiling the source data. Understanding the nature of the data that is coming is very important. We must ask, and be able to answer, the following kinds of questions at the end of the analysis:

- Does the source data contain any junk characters that need to be removed?
- Does it contain any special characters?
- Does the source data contain non-English characters (such as French or German)?
- Do any numeric columns contain null values? Which can or cannot be nullable columns?
- Is there something unique with which we can determine each record?

And the list goes on.

To analyze the source data, we should run a data profiling tool such as Talend Open Studio, DataCleaner, or AWS Glue DataBrew to analyze and visualize various metrics of the data. This activity helps us understand the data better.

Here, we will analyze the CSV data file that we need to ingest for our use case using the DataCleaner tool. Follow these steps:

1. First, you can download DataCleaner Community Edition by going to `https://datacleaner.github.io/downloads`.
2. Then, unzip the downloaded ZIP file in the desired installation folder. Based on your operating system, you can start DataCleaner using either the `datacleaner.sh` command or the `datacleaner.cmd` file present under the root installation folder. You will see a home screen, as shown in the following screenshot. Here, you can start a new data profiling job by clicking the **Build new job** button:

Figure 4.1 – DataCleaner welcome screen

3.   Then, a dialog will pop up, where you can select the data store, as shown in the following
screenshot. Here, we will browse for and select our input file called `inputData.csv`:

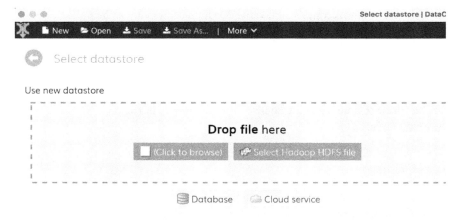

Figure 4.2 – DataCleaner – the Select datastore popup

Once the datastore is selected, we will see our data source at the top of the left pane. We should be able to see the column names of our CSV as well.

4. Now, we will drag and drop our data source `inputData.csv` file to the right pane, which is the pipeline building canvas. To profile the data, DataCleaner provides various analyzer tools under the **Analyze** menu, which is visible in the left pane. For our use case, we will be using **String analyzer**:

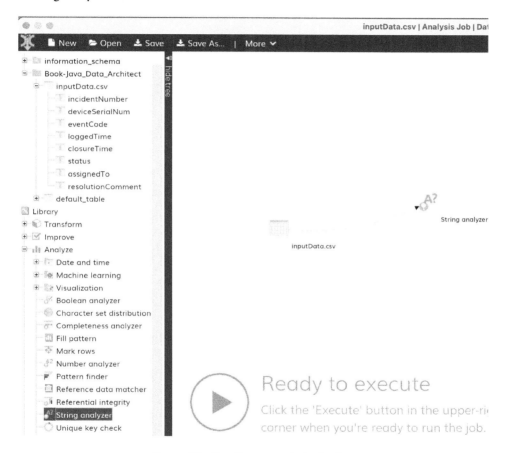

Figure 4.3 – Creating an analysis pipeline

String analyzer analyzes various string-related metrics such as the NULL count, blank count, white spaces, character case, and so on. The following screenshot shows the various configuration options of a **String analyzer**:

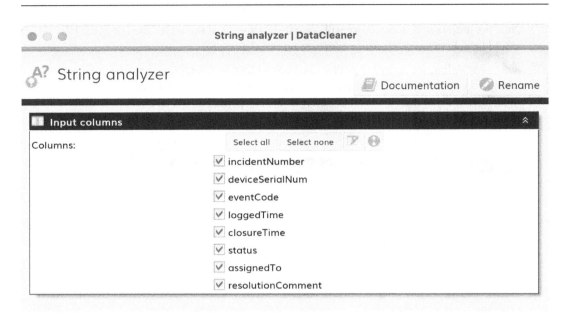

Figure 4.4 – Adding String analyzer

5. We will add another analyzer called **Completeness analyzer** to our flow to check whether any rows cannot be ingested. Each event log record must have an incidentNumber, deviceSerialNum, eventCode, and loggedTime to be an eligible entry for our data warehouse.

    If any of this information is missing, such a record will not add value to the problem that we are trying to solve. Here, **Completeness analyzer** will help us determine whether we need special checks to handle these constraints and drop records if these fields are blank. The following screenshot shows the various configuration options of **Completeness analyzer**:

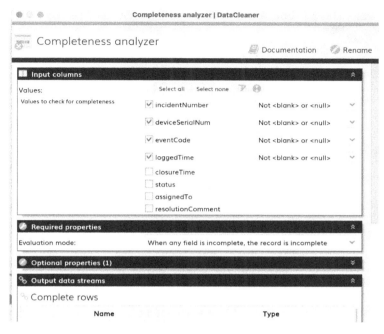

Figure 4.5 – Adding the Completeness analyzer

The final profiling pipeline for our use case can be seen in the following screenshot:

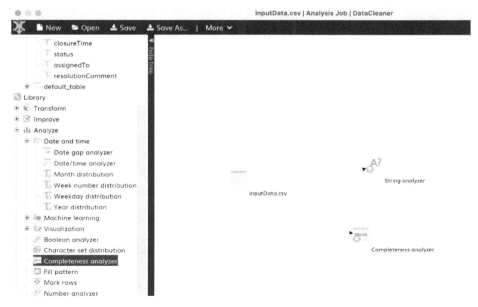

Figure 4.6 – Final analyzer pipeline

6. Once we execute this pipeline, **Analysis results** will be generated, as shown in the following screenshot:

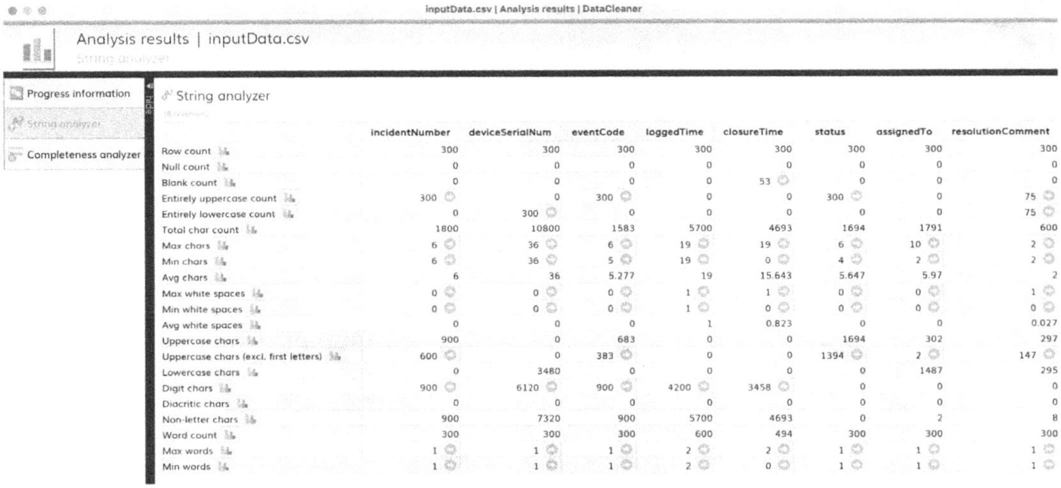

	incidentNumber	deviceSerialNum	eventCode	loggedTime	closureTime	status	assignedTo	resolutionComment
Row count	300	300	300	300	300	300	300	300
Null count	0	0	0	0	0	0	0	0
Blank count	0	0	0	0	53	0	0	0
Entirely uppercase count	300	0	300	0	0	300	0	75
Entirely lowercase count	0	300	0	0	0	0	0	75
Total char count	1800	10800	1583	5700	4693	1694	1791	600
Max chars	6	36	6	19	19	6	10	2
Min chars	6	36	5	19	0	4	2	2
Avg chars	6	36	5.277	19	15.643	5.647	5.97	2
Max white spaces	0	0	0	1	1	0	0	1
Min white spaces	0	0	0	1	0	0	0	0
Avg white spaces	0	0	0	1	0.823	0	0	0.027
Uppercase chars	900	0	683	0	0	1694	302	297
Uppercase chars (excl. first letters)	600	0	383	0	0	1394	2	147
Lowercase chars	0	3480	0	0	0	0	1487	295
Digit chars	900	6120	900	4200	3458	0	0	0
Diacritic chars	0	0	0	0	0	0	0	0
Non-letter chars	900	7320	900	5700	4693	0	2	8
Word count	300	300	300	600	494	300	300	300
Max words	1	1	1	2	2	1	1	1
Min words	1	1	1	2	0	1	1	1

Figure 4.7 – Analysis results

Such data profiling can provide us with various pieces of information about the data, which can help us adjust our tools, technologies, and transformations to create an effective and successful data engineering solution. As shown in the preceding screenshot, we can infer that the total data size is 300 rows. Out of these, 53 are open incidents. The resolution comments can have spaces in them, all `deviceSerialNum` values are lowercase, and `status` values are uppercase. Such information helps us make effective decisions while designing the solution.

For the brevity of this discussion, we are only showing one form of data profiling for a source data file. However, we can do the same for other kinds of datasets. In this use case, you can do similar data profiling for the data in the `device_dm.csv` and `event_dm.csv` files.

Now that we have understood the requirements and have a fair idea of the source data, in the next section, we will discuss how to design the model so that it can store the ingested data.

Building an effective data model

From our previous discussion and after analyzing the data, we have concluded that our data is structured, so it's suitable for being stored in a relational data model. From the requirements, we have gathered that our final data store should be a data warehouse. Keeping these two basic factors in mind, let's learn about relational data warehouse schemas.

Relational data warehouse schemas

Let's explore the popular relational data warehouse schemas that we can consider when creating our data model:

- **Star schema**: This is the most popular data warehouse schema type. As shown in the following diagram, there is a **Fact Table** in the middle where each record represents a fact or an event that has happened over time:

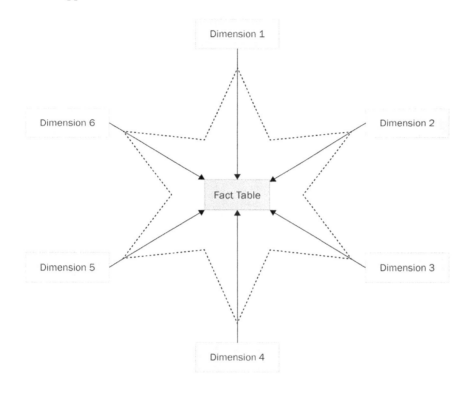

Figure 4.8 – Star schema

This **Fact Table** consists of various dimensions whose details need to be looked up from associated lookup tables called **dimension tables**. This **Fact Table** is associated with each dimension table using a foreign key. The preceding diagram shows what a star schema looks like. Since there is a **Fact Table** in the middle surrounded by multiple dimension tables on various sides, its structure looks like a star, hence its name.

- **Snowflake schema**: This is an extension of the star schema. Just like the star schema, here, there is a **Fact Table** in the middle and multiple dimension tables around it. However, in the snowflake schema, each dimension table further references other child dimension tables, making the structure look like a snowflake:

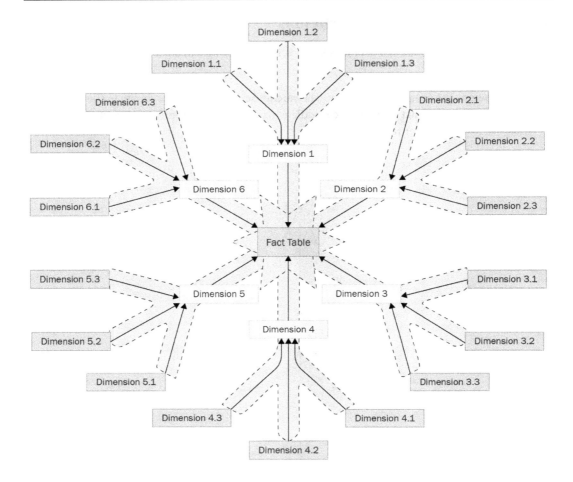

Figure 4.9 – Snowflake schema

Here, we can see how each dimension table is connected to the child dimension tables using foreign key relations, making the structure look like a snowflake, hence its name.

- **Galaxy schema**: A galaxy schema is a schema that consists of more than one fact table. Here, one or more dimension tables are shared by multiple fact tables. This schema can be visualized as a collection of two or more star schemas, hence its name.

Evaluation of the schema design

For our use case, we need to evaluate what schema design best fits our use case.

The first question that we should ask is, *In our use case, do we need multiple fact tables?* Since our fact table only consists of device events or incidents, we can only have one fact table. This eliminates any chance of having a galaxy schema as our candidate data model. Now, we must determine whether a star schema or a snowflake schema is suitable for our use case.

To choose between those two alternatives, let's look at the following columns of our `inputData.csv` file:

- `incidentNumber`
- `deviceSerialNo`
- `eventCode`
- `loggedTime`
- `closureTime`
- `status`
- `assignedTo`
- `resolutionComments`

By looking at the column names for this file, we can say that this is the device event log file. This implies that the data from the `inputData.csv` file needs to be ingested into our central fact table. But first, we need to determine whether we need to reference only dimension tables, which are complete in themselves, or whether our dimensions table needs to do further lookups in another set of dimension tables.

Let's begin by figuring out the candidate dimensions from the dataset present in `inputData.csv`. It is important to remember that candidate dimensions are determined by the purpose or goal of building the data warehouse. The purpose of the data warehouse we are building is to obtain metrics on `eventType`, device over different time intervals such as hourly, monthly, and quarterly, and closure duration metrics.

In our case, `deviceSerialNo` and `eventCode` can correspond to two dimensions called **device** and **event**. `incidentNumber` will vary in each fact record, so it's not a candidate for dimension. `status`, `loggedTime`, and `closureTime` will vary from record to record, so they are best suited for being facts and not dimensions. Since we are not doing any analysis on the `assignedTo` and `resolutionComment` fields, we can ignore those columns in our data model. In a real-world scenario, usually, incoming source data files contain hundreds of columns. However, only a small percentage of those columns are useful for solving a problem.

It is always advised to ingest only the columns that you need. This saves space, complexity, and money (remember that a lot of solutions these days are deployed on cloud platforms or are candidates for future migration to cloud platforms, and cloud platforms follow the principle of pay for what you use, so you should only ingest data that you intend to use). Apart from these, our requirements need us to mark every event on an hourly, monthly, and quarterly basis so that aggregations can easily be run on these intervals and hourly, monthly, and quarterly patterns can be analyzed. This interval tagging can be derived from `loggedTime` while saving the record. However, `hour`, `month`, and `quarter` can be stored as derived dimensions associated with our central fact table.

Hence, from our analysis, it is clear that our fact table only references those dimension tables that are complete in themselves. So, we can conclude that we will be using a star schema for our data modeling with the following set of tables:

- `DEVICE_EVENT_LOG_FACT`: This is the centralized fact table, which consists of each incident entry
- `DEVICE_DIMENSION`: This is the dimension table, which consists of device lookup data
- `EVENT_DIMENSION`: This is the dimension table, which consists of event lookup data
- `HOUR_DIMENSION`: This is the dimension table, which consists of static hour lookup data
- `MONTH_DIMENSION`: This is the dimension table, which consists of static month lookup data
- `QAURTER_DIMENSION`: This is the dimension table, which consists of static quarter lookup data

The following diagram depicts the detailed star schema data model of the data warehouse that we are building:

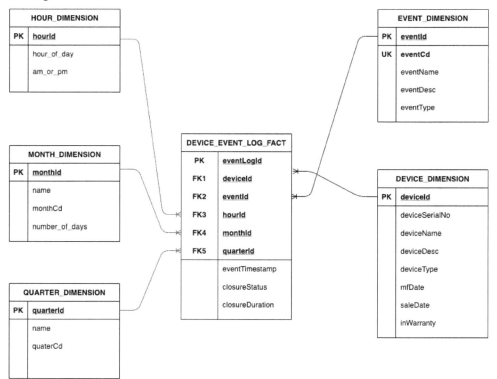

Figure 4.10 – Data model for our data warehouse

Now, let's understand the tables and their columns shown in the preceding diagram:

- In the DEVICE_EVENT_LOG_FACT table, the following is happening:

  A. We use eventLogId as the primary key, which maps to incidentNumber from our file

  B. We have foreign key fields for the DEVICE_DIMENSION, EVENT_DIMENSION, HOUR_ DIMENSION, MONTH_DIMENSION, and QUARTER_DIMENSION tables

  C. eventTimestamp, closurestatus, and closureDuration are all facts for each row in the fact table

- The columns of DEVICE_DIMENSION and EVENT_DIMENSION are determined by the need as well as the data/attributes available for the device and events in the input files – that is, device_dm.csv and event_dm.csv. However, the primary keys of these two tables (deviceId and eventId) should be system-generated sequence numbers that are assigned to a record. The primary key in these two tables is the reference column for the foreign key relationship with the fact table.

- Apart from the device and event, we have designed three other dimension tables denoting hours of the day (HOUR_DIMENSION), month (MONTH_DIMENSION), and quarter (QUARTER_ DIMENSION) of the year. These are static lookup tables, and their data will always remain constant over time.

The next design decision that needs to be made in terms of the data model is the decision to choose a database. Various **Relational Database Management Systems** (**RDBMSs**) are well suited for a data warehouse, such as Snowflake, AWS Redshift, PostgreSQL, and Oracle. While the first two options are cloud-based data warehouses, the other two options can be run both on-premises and in the cloud. For our use case, we should choose a database that is cost-effective as well as future-compatible.

Out of these choices, we will choose PostgreSQL since it is a free database that is powerful and feature-rich to host a data warehouse. Also, our application may be migrated to the cloud in the future. In that case, it can easily be migrated to AWS Redshift, as AWS Redshift is based on industry-standard PostgreSQL.

Now that we have designed our data model and chosen our database, let's go ahead and architect the solution.

Designing the solution

To design the solution for the current problem statement, let's analyze the data points or facts that are available to us right now:

- The current problem is a batch-based data engineering problem

- The problem at hand is a data ingestion problem

- Our source is CSV files containing structured data

- Our target is a PostgreSQL data warehouse

- Our data warehouse follows a star schema, with one fact table, two dynamic dimension tables, and three static dimension tables

- We should choose a technology that is independent of the deployment platform, considering that our solution can be migrated to the cloud in the future

- For the context and scope of this book, we will explore optimum solutions based on Java-based technologies

Based on the preceding facts, we can conclude that we have to build three similar data ingestion pipelines – one for the fact table and two others for the dynamic dimension tables. At this point, we must ask ourselves, *What happens to the file if the file ingestion is successful or if it fails? How do we avoid reading the file again?*

We will read the file from the input folder and ingest it into the data warehouse. If it fails, we will move the file to an error folder; otherwise, we will move it to an archive folder. The following diagram shows our findings and provides an overview of our proposed solution:

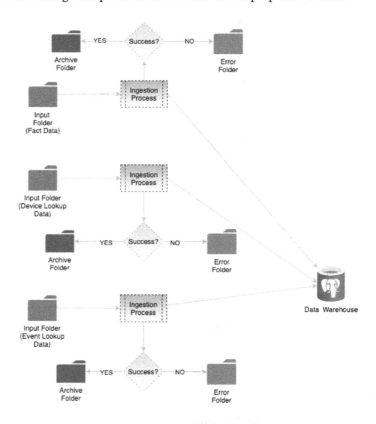

Figure – 4.11 – Solution overview

However, this proposed solution is a 10,000-feet view of it. There are still many questions that have been left unanswered in this solution. For instance, there are no details about the ingestion processes or the technology that we should use to solve this problem.

First, let's try to decide on a technology based on the facts we have at hand. We need to find a Java-based ETL technology that supports batch ingestion. Also, it should have easy-to-use JDBC support to write and read data from PostgreSQL. We also need to have a scheduler to schedule the batch ingestion job and should have a retry ability mechanism. Also, our data is not huge, so we want to avoid big data-based ETL tools.

Spring Batch fits all these requirements. Spring Batch is an excellent Java-based ETL tool for building batch jobs. It comes with a job scheduler and a job repository. Also, since it is a part of the Spring Framework, it can easily be integrated with various tools and technologies with Spring Boot and Spring integration. The following diagram shows the high-level components of the Spring Batch architecture:

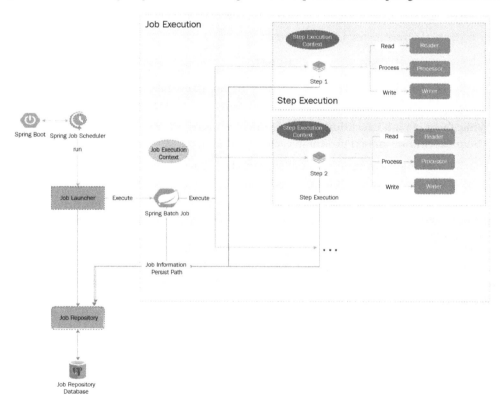

Figure 4.12 – Spring Batch architecture

The preceding diagram denotes how a Spring Batch job works. Let's look at the various steps that a Spring Batch job goes through to be executed:

1. The Spring Batch job uses Spring's job scheduler to schedule a job.

2. **Spring Job Scheduler** runs **Job Launcher**, which, in turn, executes a **Spring Batch job**. It also creates a job instance at this point and persists this information in the job repository database.

3. The **Spring Batch job** tracks all job-related information in a job repository database automatically using batch execution tables. The various batch execution tables are as follows:

    *   `batch_job_instance`

    *   `batch_job_execution`

    *   `batch_job_execution_params`

    *   `batch_step_execution`

    *   `batch_job_execution_context`

    *   `batch_step_execution_context`

4. The **Spring Batch job**, which is executed by the **Job Launcher**, initiates individual steps to perform the job. Each step performs a specific task to achieve the overall aim of the job.

5. While there is a **Job Execution Context** present across all the steps of a job instance, there is a **Step Execution Context** present in each execution step.

6. Usually, a Spring Batch configuration helps stitch together each step in the desired sequence to create the Spring Batch pipeline.

7. Each step, in turn, reads the data using `Reader` or `ItemReader`, processes the data using `Processor` or `ItemProcessor`, and writes the processed data using `Writer` or `ItemWriter`.

Now that we have a fair understanding of the Spring Batch architecture, we will architect our ingestion pipeline using the Spring Batch job framework. The following diagram shows the architecture of our data ingestion pipeline:

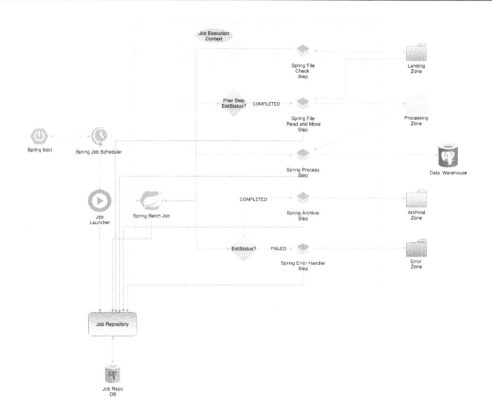

Figure 4.13 – Solution architecture

Let's look at this solution in more detail:

1. Like all **Spring Batch jobs**, the **Spring Job Scheduler** schedules a **Job Launcher**, which instantiates the Spring job.

2. In our use case, we will use a total of three sequential and two conditional steps to complete the job.

3. In the first steps, the application looks at whether there is a new file in the input folder or **Landing Zone**. If there is, it updates the file path in `JobExecutionContext` and marks **ExitStatus** as **COMPLETED**.

4. If **ExitStatus** of the first step is **COMPLETED**, then the second step is initiated. The second step moves the file mentioned in the file path (of `JobExecutionContext`) from **Landing Zone** to **Processing Zone**.

5. Upon completing the second step, the third step (**Spring Process Step**) is initiated. The third step transforms and loads the data into a data warehouse.

6. Upon completing the third step, **Spring Archival Step** is initiated, which moves the processed file from the process folder to the archive folder.

7. However, if **String Process Step** fails, **Spring Error Handler Step** will be initiated, where it moves the file from the **Processing Zone** folder to the **Error Zone** folder.

In this section, we learned how to logically divide the solution using the available facts and data points and come up with an optimal architecture for the problem. We also learned how the effectiveness of each solution is dependent on the technology stack that we choose.

In the next section, we will learn how to implement and test our solution using Spring Batch and related technologies.

Implementing and unit testing the solution

In this section, we will build the Spring Batch application to implement the solution that we designed in the preceding section. We will also run and test the solution.

First, we must understand that different jobs will have their own schedules. However, the dimension tables need to be loaded before the fact table, because the dimension tables are the lookup tables.

For the brevity of our discussion, we will only implement the Spring Batch application for the fact table. In this implementation, we will load the device data and event data from CSV to the table manually. However, you can follow the lead of the discussion by implementing the solution and developing two different Spring Batch applications for the device and event dimension tables. In this implementation, we will assume that the device and event data have already been loaded into the data warehouse.

You can do that manually by executing the DMLs present at the following GitHub link: `https://github.com/PacktPublishing/Scalable-Data-Architecture-with-Java/blob/main/Chapter04/SQL/chapter4_ddl_dml.sql`.

We need to begin by creating a Spring Boot Maven project and adding the required Maven dependencies. The following Maven dependencies should be added to the `pom.xml` file, as follows:

```
<dependencies>
    <dependency>
        <groupId>org.springframework.boot</groupId>
        <artifactId>spring-boot-starter-batch</artifactId>
        <version>2.4.0</version>
    </dependency>
    <dependency>
        <groupId>org.springframework.boot</groupId>
        <artifactId>spring-boot-starter-jdbc</artifactId>
```

```xml
                <version>2.4.0</version>
        </dependency>
        <!-- https://mvnrepository.com/artifact/org.postgresql/
postgresql -->
        <dependency>
                <groupId>org.postgresql</groupId>
                <artifactId>postgresql</artifactId>
                <version>42.3.1</version>
        </dependency>
        <dependency>
                <groupId>org.slf4j</groupId>
                <artifactId>slf4j-api</artifactId>
                <version>2.0.0-alpha0</version>
        </dependency>
        <dependency>
                <groupId>org.slf4j</groupId>
                <artifactId>slf4j-log4j12</artifactId>
                <version>2.0.0-alpha0</version>
                <scope>runtime</scope>
        </dependency>
    </dependencies>
```

Two Spring dependencies are added here: `spring-boot-starter-batch` is added for Spring Batch and `spring-boot-starter-jdbc` is added for communicating with the `postgreSQL` database (which is used as a data warehouse and the Spring Batch repository database). Apart from this, the JDBC driver for PostgreSQL and the logging dependencies are added.

As per our architecture, let's start by creating the entry point of the Spring Boot application, which is the `Main` class, and initializing the job scheduler. The following code denotes our `Main` class:

```java
@EnableConfigurationProperties
@EnableScheduling
@ComponentScan ({"com.scalabledataarchitecture.etl.*", "com.
scalabledataarchitecture.etl.config.*"})
@SpringBootApplication
public class EtlDatawarehouseApplication {

    private static Logger LOGGER = LoggerFactory.
getLogger(EtlDatawarehouseApplication.class);
```

```
@Autowired
JobLauncher jobLauncher;

@Autowired
Job etlJob;

public static void main(String[] args) {
    try {
        SpringApplication.run(EtlDatawarehouseApplication.
class, args);
    }catch (Throwable ex){
        LOGGER.error("Failed to start Spring Boot
application: ",ex);
    }
}


@Scheduled(cron = "0 */1 * * * ?")
public void perform() throws Exception
{
    JobParameters params = new JobParametersBuilder().
addString("JobID", String.valueOf(System.currentTimeMillis())).
toJobParameters();
    jobLauncher.run(etlJob, params);
}

}
```

The @SpringBootApplication annotation denotes that this class is the entry point of the Spring Boot application. Also, please note that the @EnableScheduling annotation denotes that this application supports Spring job scheduling. A method with the @Scheduled annotation helps perform the scheduled function at the configured schedule interval.

The Spring Batch job scheduler supports all of the three formats as shown in the following block of code:

```
@Scheduled(fixedDelayString = "${fixedDelay.in.milliseconds}")
@Scheduled(fixedRateString = "${fixedRate.in.milliseconds}")
@Scheduled(cron = "${cron.expression}")
```

Here, fixedDelayString makes sure that there is a delay of *n* milliseconds between the end of a job and the beginning of another job. fixedRateString runs the scheduled job every *n* milliseconds, while cron schedules the job using some cron expression. In our case, we are using a cron expression to schedule the perform() method.

The perform() method adds a job parameter called JobID and triggers a Spring Batch job called etlJob using jobLauncher. jobLauncher is an auto-wired bean of the JobLauncher type.

The etlJob field in the EtlDatawarehouseApplication class, as shown earlier, is also auto-wired and hence is a Spring bean.

Next, we will explore the Spring Batch configuration file where the etlJob bean is created:

```
@Configuration
@EnableBatchProcessing
public class BatchJobConfiguration {

    @Bean
    public Job etlJob(JobBuilderFactory jobs,
                        Step fileCheck, Step fileMoveToProcess,
Step processFile,Step fileMoveToArchive, Step fileMoveToError)
{
        return jobs.get("etlJob")
                .start(fileCheck).on(ExitStatus.STOPPED.
getExitCode()).end()
                .next(fileMoveToProcess)
                .next(processFile).on(ExitStatus.COMPLETED.
getExitCode()).to(fileMoveToArchive)
                .from(processFile).on(ExitStatus.FAILED.
getExitCode()).to(fileMoveToError)
                .end()
                .build();
    }
}
```

As you can see, the class is annotated with @Configuration and @EnableBatchProcessing. This ensures that the BatchJobConfiguration class is registered as a configuration bean in Spring, as well as a couple of other batch-related bean components, such as JobLauncher, JobBuilderFactory, JobRepository, and JobExplorer.

The etlJob() function uses JobBuilderFactory to create the step pipeline, as described during the design phase. The etlJob pipeline starts with the fileCheck step. If the exit status of the `fileCheck` step is `STOPPED`, the batch job ends; otherwise, it moves to the next step – that is, `fileMoveToProcess`. The next step is `processFile`. On returning `COMPLETED` from the `processFile` step, the `moveToArchive` step is invoked. However, on returning `ExitStatus` as `FAILED`, the `moveToError` step is invoked.

However, we can create an `etlJob` bean. To do so, we need to create all the step beans that are stitched together to form the batch job pipeline. Let's begin by looking at how to create the `fileCheck` bean.

To create the `fileCheck` bean, we have written the following two classes:

- `FileCheckConfiguration`: A configuration class where the `fileCheck` bean is initialized.
- `FileCheckingTasklet`: A Tasklet class for the `fileCheck` step. `Tasklet` is meant to perform a single task within a step.

`FileCheckingTasklet` is a `Tasklet`, so it will implement a `Tasklet` interface. The code will be similar to the following:

```
public class FileCheckingTasklet implements Tasklet{
//...
}
```

`Tasklet` contains only one method – `execute()` – that must be implemented. It has the following type signature:

```
public RepeatStatus execute(StepContribution stepContribution,
ChunkContext chunkContext) throws Exception
```

In `FileCheckingTasklet`, we wish to check whether any file is present in the landing zone or not. Our main aim for using this `Tasklet` is to change the `EXITSTATUS` property of the task based on whether the file is present or not. Spring Batch provides an interface called `StepExecutionListener` that enables us to modify `EXITSTATUS` based on our requirements. This can be done by implementing the `afterStep()` method of `StepExecutionListener`. The interface definition of `StepExecutionListener` looks as follows:

```
public interface StepExecutionListener extends StepListener {
    void beforeStep(StepExecution var1);

    @Nullable
    ExitStatus afterStep(StepExecution var1);
}
```

So, our `FileCheckingTasklet` will look similar to the following:

```
public class FileCheckingTasklet implements Tasklet,
```

```
StepExecutionListener {
//...
@Override
public RepeatStatus execute(StepContribution stepContribution,
ChunkContext chunkContext) throws Exception {
//...
}
@Override
public ExitStatus afterStep(StepExecution stepExecution) {
//...
}
}
```

Now, let's understand the logic that we want to execute in this `Tasklet`. We want to list all the files in the landing zone directory. If no files are present, we want to set EXITSTATUS to STOPPED. If we find one or more files, we want to set EXITSTATUS to COMPLETED. If an error occurs while listing the directory, we will set EXITSTATUS to FAILED. Since we can modify EXITSTATUS in the `afterStep()` method, we will write our logic in that method. However, we want to configure our landing zone folder in our application. We can do that by using a configuration POJO called `EnvFolderProperty` (we will discuss the code of this class later in this chapter). Here is the logic of the `afterstep()` method:

```
@Override
public ExitStatus afterStep(StepExecution stepExecution) {
    Path dir = Paths.get(envFolderProperty.getRead());
    LOGGER.debug("Checking if read directory {} contains some
files...", dir);
    try {
        List<Path> files = Files.list(dir).filter(p -> !Files.
isDirectory(p)).collect(Collectors.toList());
        if(files.isEmpty()) {
            LOGGER.info("Read directory {} does not contain any
file. The job is stopped.", dir);
            return ExitStatus.STOPPED;
        }
        LOGGER.info("Read directory {} is not empty. We
continue the job.", dir);
        return ExitStatus.COMPLETED;
    } catch (IOException e) {
```

```
        LOGGER.error("An error occured while checking if read
directory contains files.", e);
        return ExitStatus.FAILED;
    }
}
```

Since we don't want to do any other processing in this `Tasklet`, we will let the `execute()` method pass with a `RepeatStatus` of `FINISHED`. So, our full code for `FileCheckingTasklet` will look as follows:

```
public class FileCheckingTasklet implements Tasklet,
StepExecutionListener {
    private final static Logger LOGGER = LoggerFactory.
getLogger(FileCheckingTasklet.class);

    private final EnvFolderProperty envFolderProperty;

    public FileCheckingTasklet(EnvFolderProperty
envFolderProperty) {
        this.envFolderProperty = envFolderProperty;
    }

    @Override
    public void beforeStep(StepExecution stepExecution) {
    }

    @Override
    public RepeatStatus execute(StepContribution
stepContribution, ChunkContext chunkContext) throws Exception {
        return RepeatStatus.FINISHED;
    }

    @Override
    public ExitStatus afterStep(StepExecution stepExecution) {
        // Source code as shown in previous discussion ...
    }
}
```

Now, let's see how we can use `FileCheckingTasklet` to create the `fileCheck` step. In the `FileCheckConfiguration` configuration class, first, we create a bean for `FileCheckingTasklet`, as follows:

```
@Bean
public Tasklet fileCheckingTasklet(EnvFolderProperty
envFolderProperty) {
    return new FileCheckingTasklet(envFolderProperty);
}
```

Then, we use this bean to create the `fileCheck` step bean, as follows:

```
@Bean
public Step fileCheck(StepBuilderFactory stepBuilderFactory,
Tasklet fileCheckingTasklet) {
    return stepBuilderFactory.get("fileCheck")
            .tasklet(fileCheckingTasklet)
            .build();
}
```

Finally, the full code for the `FileCheckConfiguration` configuration class will look as follows:

```
@Configuration
public class FileCheckConfiguration {

    @Bean
    public Tasklet fileCheckingTasklet(EnvFolderProperty
envFolderProperty) {
        return new FileCheckingTasklet(envFolderProperty);
    }

    @Bean
    public Step fileCheck(StepBuilderFactory
stepBuilderFactory, Tasklet fileCheckingTasklet) {
        return stepBuilderFactory.get("fileCheck")
                .tasklet(fileCheckingTasklet)
                .build();
    }
}
```

In the preceding steps, we learned how to create a step using `Tasklet` and the `StepExecutionListener` interface and instantiate and utilize them using Spring's powerful annotations, such as `@Bean`, `@Configuration`, and `@AutoWired`.

> **Important note**
> Spring Batch provides various listeners (listener interfaces) to intercept, listen, and react to the Spring Batch job flow at different levels. If you are interested, you can learn more about Spring Batch listeners at `https://howtodoinjava.com/spring-batch/spring-batch-event-listeners/`.

Now, we will move on to the next step, called `moveFileToProcess`, and see how we can implement its design. Again, to implement the `moveFileToProcess` step, we will be writing a configuration file called `FileMoveToProcessConfiguration` and a tasklet file called `FileMoveToProcessTasklet`.

First, let's build our `FileMoveToProcessTasklet` tasklet. To build our tasklet, we will define the tasks that we want to achieve while using it. Here are the tasks that we want to accomplish using this tasklet:

- List the files present in the landing zone

- Move one file at a time to the process zone

- Add the destination full file path (the file path in the processing zone) as the key-value entry to `JobExecutionContext`

Just like the previous tasklet that we developed, `FileMoveToProcessTasklet` will also implement the `Tasklet` and `StepExecutionListener` interfaces.

The following code shows our implementation of the `execute()` function of the `Tasklet` interface:

```
@Override
public RepeatStatus execute(StepContribution stepContribution,
ChunkContext chunkContext) throws Exception {
    Path dir = Paths.get(envFolderProperty.getRead());
    assert Files.isDirectory(dir);
    List<Path> files = Files.list(dir).filter(p -> !Files.
isDirectory(p)).collect(Collectors.toList());
    if(!files.isEmpty()) {
        Path file = files.get(0);
        Path dest = Paths.get(envFolderProperty.getProcess() +
File.separator + file.getFileName());
        LOGGER.info("Moving {} to {}", file, dest);
```

```
        Files.move(file, dest);
        filepath = dest;
    }
    return RepeatStatus.FINISHED;
}
```

First, we list the files from the landing zone (the read directory path) and if the list of files is not empty, we get the first file and move it to the destination path. Here, we create the destination path by appending the filename and file separator to the processing directory.

Once the file is successfully moved to the destination, we set the value of the `filepath` instance variable as the destination path where the file has been moved. We will use this in our implementation of the `afterStep()` method. Now, let's look at the implementation of the `afterStep()` method, as follows:

```
@Override
public ExitStatus afterStep(StepExecution stepExecution) {
    if(filepath != null) {
        stepExecution.getJobExecution().getExecutionContext().
put("filepath", filepath);
        stepExecution.getJobExecution().getExecutionContext().
put("filepathName", filepath.toString());
    }
    return ExitStatus.COMPLETED;
}
```

In the `afterStep()` method implementation, we store two key-value entries (`filePath` and `filePathName`) in `JobExecutionContext` if `filePath` is not null (which means at least one file was present in the landing zone during tasklet execution and has been successfully moved to the processing zone by the tasklet).

Now, let's see the full code for the `FileMoveToProcessTasklet` class:

```
public class FileMoveToProcessTasklet implements Tasklet,
StepExecutionListener {
    private final static Logger LOGGER = LoggerFactory.
getLogger(FileMoveToProcessTasklet.class);
    private final EnvFolderProperty envFolderProperty;
    private Path filepath;

    public FileMoveToProcessTasklet(EnvFolderProperty
```

```
envFolderProperty) {
        this.envFolderProperty = envFolderProperty;
    }

    @Override
    public void beforeStep(StepExecution stepExecution) {
    }

    @Override
    public RepeatStatus execute(StepContribution
stepContribution, ChunkContext chunkContext) throws Exception {
    // Source code as shown in previous discussion ...
  }

    @Override
    public ExitStatus afterStep(StepExecution stepExecution) {
    // Source code as shown in previous discussion ...
    }

}
```

The source code of `FileMoveToProcessConfiguration` will be very similar to `FileCheckConfiguration`, which we discussed earlier.

The source code of `FileMoveToProcessConfiguration` is as follows:

```
@Configuration
public class FileMoveToProcessConfiguration {

    @Bean
    public Tasklet fileMoveToProcessTasklet(EnvFolderProperty
envFolderProperty) {
        return new FileMoveToProcessTasklet(envFolderProperty);
    }

    @Bean
    public Step fileMoveToProcess(StepBuilderFactory
stepBuilderFactory, Tasklet fileMoveToProcessTasklet) {
```

```
        return stepBuilderFactory.get("fileMoveToProcess")
                .tasklet(fileMoveToProcessTasklet)
                .build();
    }
}
```

Now, we will learn how to develop the `processFile` step. This is an important step as all the transformation and ingestion happens here. This step follows a typical `SpringBatch` step template, where there is an `ItemReader`, `ItemProcessor`, and `ItemWriter`. They are stitched together to form the step pipeline.

First, let's look at the source code for building the step pipeline in the `processFile()` method of the `ProcessFileConfiguration` configuration class:

```
@Bean
public Step processFile(StepBuilderFactory
stepBuilderFactory, ItemReader<EventLogODL> csvRecordReader,
JdbcBatchItemWriter<DeviceEventLogFact> jdbcWriter) {
    return stepBuilderFactory.get("processFile")
            .<EventLogODL, DeviceEventLogFact>chunk(chunkSize)
            .reader(csvRecordReader)
            .processor(deviceEventProcessor)
            .writer(jdbcWriter)
            .build();
}
```

Here, we are building the step from an `ItemReader` bean called `csvRecordReader`, which reads the records from the CSV file and returns a set of `EventLogODL` POJO objects. An `ItemProcessor` bean called `deviceEventProcessor`, which reads each `EventLogODL` POJO and transforms them into `DeviceEventLogFact` POJOs, and an `ItemWriter` bean called `jdbcWriter`, which reads each record as a `DeviceEventLogFact` POJO and persists them in the PostgreSQL data warehouse. We also mention `chunk` while building the pipeline while using `chunkSize` as a configurable parameter (for learning purposes, we will test with a `chunkSize` of 1).

Before we explain how to develop the `ItemReader` bean, let's look at the source code of the `EventLogODL` POJO class:

```
public class EventLogODL {
    private String incidentNumber;
    private String deviceSerialNum;
    private String eventCode;
```

```
    private String loggedTime;
    private String closureTime;
    private String status;
    private String assignedTo;
    private String resolutionComment;


  // Getters and Setter of the instance fields


}
```

Now, let's look at the method that creates the `ItemReader` bean:

```
@Bean
@StepScope
public FlatFileItemReader<EventLogODL> csvRecordReader(@
Value("#{jobExecutionContext['filepathName']}") String
filePathName)
        throws UnexpectedInputException, ParseException {
    FlatFileItemReader<EventLogODL> reader = new
FlatFileItemReader<EventLogODL>();
    DelimitedLineTokenizer tokenizer = new
DelimitedLineTokenizer();
    String[] tokens = { "incidentNumber","deviceSerialNum",
"eventCode","loggedTime","closureTime","status","assignedTo",
"resolutionComment" };
    tokenizer.setNames(tokens);
    reader.setResource(new FileSystemResource(filePathName));
    reader.setLinesToSkip(1);
    DefaultLineMapper<EventLogODL> lineMapper =
            new DefaultLineMapper<EventLogODL>();
    lineMapper.setLineTokenizer(tokenizer);
    lineMapper.setFieldSetMapper(new
BeanWrapperFieldSetMapper<EventLogODL>() {
        {
            setTargetType(EventLogODL.class);
        }
    });
    reader.setLineMapper(lineMapper);
    return reader;
```

```
}
```

Here, we are using Spring Batch's inbuilt `FlatFileItemReader` to read the CSV source file. Since we need to dynamically read `filePathName` from `jobExecutionContext`, which we set in the previous step, we used the `@SetScope` annotation, which changes the default bean scope from a singleton to a step-specific object. This annotation is especially helpful for late binding where we want to read some parameters dynamically from `JobExecutionContext` or `StepExecutionContext`. Also, we are creating a delimited tokenizer with `fieldNames` and a `BeanWrapperFieldSetMapper` to map each record to the `EventLogODL` POJO and set the corresponding properties of the `FlatfileItemReader` instance.

Troubleshooting tips

In an ideal world, all data is perfect, and our job should run fine every time. But we don't live in an ideal world. What happens if the data is corrupted? What happens if a few records in the file are not following the proper schema? How do we handle such situations?

There is no simple answer. However, Spring Batch gives you few capabilities to handle failures. If the file itself is corrupted and not readable or the file doesn't have proper read and write permissions, then it will go to a **Step Failure**, which results in moving the file to the error folder (as discussed earlier in this chapter). However, some exceptions are encountered while processing, which should not result in **Step Failure** (for example a parsing error for a few lines should not result in the whole file being discarded) but should be skipped instead. Such scenarios can be handled by setting a `faultTolerance` during building the step. This can either be done by using the `skipLimit()`, `skip()`, and `noSkip()` methods or using a custom `SkipPolicy`. In our example, we can add a fault tolerance to the `processFile` method of the `ProcessFileConfiguration` class and skip certain kinds of exceptions while ensuring few other types of exceptions cause a step failure. An example is shown in the following code:

```
return stepBuilderFactory.get("processFile")
        .<EventLogODL, DeviceEventLogFact>chunk(chunkSize)
        .reader(csvRecordReader)
.faultTolerant().skipLimit(20).skip(SAXException.class).
noSkip(AccessDeniedException.class)
        .processor(deviceEventProcessor)
        .writer(jdbcWriter)
        .build();
```

As we can see, we can add fault tolerance by chaining the `faultTolerant()` method in `stepBuilderFactory.build()`. Then, we can chain the `skip()` method so that it skips 20 errors of the `SAXException` type and use the `noSkip()` method to ensure `AccessDeniedException` will always cause a **Step Failure**.

Now, let's see how we can develop our custom `ItemProcessor`. The source code of the custom `ItemProcessor`, called `DeviceEventProcessor`, is shown in the following block:

```
@Component
public class DeviceEventProcessor implements
ItemProcessor<EventLogODL, DeviceEventLogFact> {

    @Autowired
    DeviceEventLogMapper deviceEventLogMapper;


    @Override
    public DeviceEventLogFact process(EventLogODL eventLogODL)
throws Exception {
        return deviceEventLogMapper.map(eventLogODL);
    }
}
```

As you can see, we have implemented the `ItemProcessor` interface, where we need to implement the `process()` method. To convert the `EventLogODL` POJO into a `DeviceEventLogFact` POJO, we have created a delegate component called `DeviceEventLogMapper`. The source code of `DeviceEventLogMapper` is as follows:

```
@Component
public class DeviceEventLogMapper {

    @Autowired
    JdbcTemplate jdbcTemplate;

    public DeviceEventLogFact map(EventLogODL eventLogODL){
        String sqlForMapper = createQuery(eventLogODL);
        return jdbcTemplate.queryForObject(sqlForMapper,new
BeanPropertyRowMapper<>(DeviceEventLogFact.class));
    }


    private String createQuery(EventLogODL eventLogODL){
        return String.format("WITH DEVICE_DM AS\n" +
                "\t(SELECT '%s' AS eventLogId "+
...,eventLogODL.getIncidentNumber(),eventLogODL.
```

```
getDeviceSerialNum(),...);
    }
```

Since we are developing the code for the fact table, we need various primary keys to be fetched from different dimension tables to populate our `DeviceEventLogFact` POJO. Here, we are dynamically creating a query by using `jdbcTemplate` to fetch the dimension primary keys from the data warehouse and populating the `DeviceEventLogFact` POJO from its `resultset`. The complete source code for `DeviceEventLogMapper` is available on GitHub at `https://github.com/PacktPublishing/Scalable-Data-Architecture-with-Java/blob/main/Chapter04/SpringBatchApp/EtlDatawarehouse/src/main/java/com/scalabledataarchitecture/etl/steps/DeviceEventLogMapper.java`.

Finally, we will create an `ItemWriter` called `jdbcwriter` in the `ProcessFileConfiguration` class, as follows:

```
@Bean
public JdbcBatchItemWriter<DeviceEventLogFact> jdbcWriter() {
    JdbcBatchItemWriter<DeviceEventLogFact> writer = new
JdbcBatchItemWriter<>();
    writer.setItemSqlParameterSourceProvider(new
BeanPropertyItemSqlParameterSourceProvider<>());
    writer.setSql("INSERT INTO chapter4.device_event_log_
fact(eventlogid,deviceid,eventid,hourid,monthid,quarterid,
eventtimestamp,closurestatus,closureduration)
VALUES (:eventLogId, :deviceId, :eventId, :hourId,
:monthId, :quarterId, :eventTimestamp, :closureStatus,
:closureDuration)");
    writer.setDataSource(this.dataSource);
    return writer;
}
```

Finally, our source code for the `ProcessFileConfiguration` class looks as follows:

```
@Configuration
@SuppressWarnings("SpringJavaAutowiringInspection")
public class ProcessFileConfiguration {
    private final static Logger LOGGER = LoggerFactory.
getLogger(ProcessFileConfiguration.class);

    @Value("${process.chunk_size:1}")
    int chunkSize;
```

```java
    @Autowired
    DataSource dataSource;

    @Autowired
    DeviceEventProcessor deviceEventProcessor;

    @Bean
    @StepScope
    public FlatFileItemReader<EventLogODL> csvRecordReader(@
Value("#{jobExecutionContext['filepathName']}")  String
filePathName)
            throws UnexpectedInputException, ParseException {
        // Source code as shown in previous discussion ...
    }

    @Bean
    public JdbcBatchItemWriter<DeviceEventLogFact> jdbcWriter()
{

// Source code as shown in previous discussion ...

  }

    @Bean
    public Step processFile(StepBuilderFactory
stepBuilderFactory, ItemReader<EventLogODL> csvRecordReader,
JdbcBatchItemWriter<DeviceEventLogFact> jdbcWriter) {

    // Source code as shown in previous discussion ...
    }
}
```

Now that we have built our code, we will configure our properties in the application.yaml file, which is present in the resource folder, as follows:

```yaml
spring:
  datasource:
```

```
  url: jdbc:postgresql://localhost:5432/sinchan
  username: postgres
  driverClassName: org.postgresql.Driver
batch:
  initialize-schema: always
env:
  folder:
    read: /Users/sinchan/Documents/Personal_Docs/Careers/Book-
Java_Data_Architect/chapter_4_pgrm/landing
    process: /Users/sinchan/Documents/Personal_Docs/Careers/
Book-Java_Data_Architect/chapter_4_pgrm/process
    archive: /Users/sinchan/Documents/Personal_Docs/Careers/
Book-Java_Data_Architect/chapter_4_pgrm/archive
    error: /Users/sinchan/Documents/Personal_Docs/Careers/Book-
Java_Data_Architect/chapter_4_pgrm/error
```

As shown in the `.yaml` file, we have to mention the `spring.datasource` properties so that Spring JDBC can automatically auto-wire the data source component.

Our final code structure will look as follows:

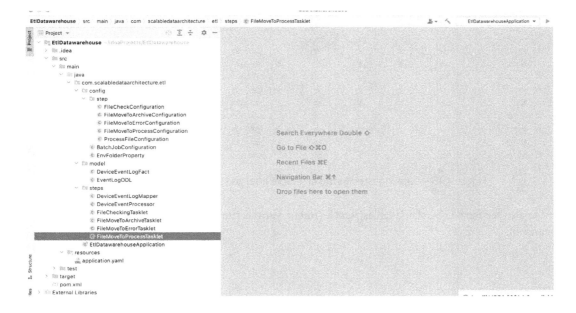

Figure 4.14 – Project structure of the code

We can run and test our program by running our Spring Boot application from our favorite IDE by running the `Main` class – that is, `EtlDataWarehouseApplication`. We must install Postgres, create the schema and the database tables, and populate all the dimension tables before we run our Spring Batch application here. Detailed run instructions can be found in this book's GitHub repository.

Once we have run our application and placed our data in the landing zone, it gets ingested into our data warehouse fact table, and the CSV files get moved to the archival zone, as shown in the following screenshot:

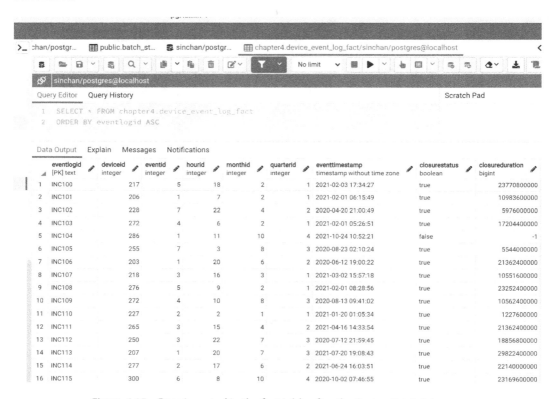

Figure 4.15 – Data ingested in the fact table after the Spring Batch job runs

We can also see the batch-related tables, which contain various run statistics, as shown in the following screenshot:

Figure 4.16 – Batch job execution log

The preceding screenshot shows the batch execution log for the different batch jobs that have run. We can learn more about a specific job by looking at the batch step execution log, as shown in the following screenshot:

step_execution_id [PK] bigint	version bigint	step_name character varying (100)	job_execution_id bigint	start_time timestamp without time zone	end_time timestamp without time zc	status character varying (1	commit_count bigint	read_cou bigint
100	3	fileCheck	57	2021-11-09 00:56:43.942	2021-11-09 00:56:43.987	COMPLETED	1	
101	3	fileCheck	58	2021-11-09 00:57:00.036	2021-11-09 00:57:00.051	COMPLETED	1	
102	3	fileCheck	59	2021-11-09 00:58:00.031	2021-11-09 00:58:00.046	COMPLETED	1	
103	3	fileMoveToProcess	59	2021-11-09 00:58:00.064	2021-11-09 00:58:00.085	COMPLETED	1	
104	303	processFile	59	2021-11-09 00:58:00.115	2021-11-09 00:58:01.056	COMPLETED	301	
105	3	fileMoveToArchive	59	2021-11-09 00:58:01.06	2021-11-09 00:58:01.063	COMPLETED	1	
106	3	fileCheck	60	2021-11-09 00:59:00.025	2021-11-09 00:59:00.034	COMPLETED	1	

Figure 4.17 – Step execution log for the batch jobs

With that, we have analyzed, architected, designed, developed, and tested a batch-based ETL data ingestion pipeline successfully. As mentioned in the *Technical requirements* section, the detailed source code is available in this book's GitHub repository.

Summary

In this chapter, we learned how to analyze a data engineering requirement from scratch, draw a definite conclusion, and extract facts that will help us in our architectural decision-making process. Next, we learned how to profile source data and how such an analysis helps us build better data engineering solutions. Going further, we used facts, requirements, and our analysis to build a robust and effective architecture for a batch-based data engineering problem with a low or medium volume of data. Finally, we mapped the design to build an effective ETL batch-based data ingestion pipeline using Spring Batch and test it. Along the way, you learned how to analyze a data engineering problem from scratch and how to build similar pipelines effectively for when you are presented with a similar problem next time around.

Now that we have successfully architected and developed a batch-based solution for medium- and low-volume data engineering problems, in the next chapter, we will learn how to build an effective data engineering solution for dealing with huge data volumes. In the next chapter, we will discuss an interesting use case for building an effective batch-based big data solution.

5

Architecting a Batch Processing Pipeline

In the previous chapter, we learned how to architect medium- to low-volume batch-based solutions using Spring Batch. We also learned how to profile such data using DataCleaner. However, with data growth becoming exponential, most companies have to deal with huge volumes of data and analyze it to their advantage.

In this chapter, we will discuss how to analyze, profile, and architect a big data solution for a batch-based pipeline. Here, we will learn how to choose the technology stack and design a data pipeline to create an optimized and cost-efficient big data solution. We will also learn how to implement this solution using Java, Spark, and various AWS components and test our solution. After that, we will discuss how to optimize the solution to be more time and cost-efficient. By the end of this chapter, you will know how to architect and implement a data analysis pipeline in AWS using S3, Apache Spark (Java), AWS EMR, AWS Lambda, and AWS Athena. You will also know how to fine-tune the code for optimized performance, as well as how to plan and optimize the cost of implementations.

In this chapter, we're going to cover the following main topics:

- Developing the architecture and choosing the right tools
- Implementing the solution
- Querying the ODL using AWS Athena

Technical requirements

To follow along with this chapter, you'll need the following:

- Prior knowledge of Java
- Prior knowledge of the basics of Apache Spark
- Java 1.8 or above installed on your local machine

- IntelliJ Idea community or ultimate edition installed on your local machine
- An AWS account

The code for this chapter can be found in this book's GitHub repository: `https://github.com/PacktPublishing/Scalable-Data-Architecture-with-Java/tree/main/Chapter05`.

Developing the architecture and choosing the right tools

In data engineering, after the data has been successfully ingested and stored in a data lake or a data warehouse, often, it needs to be mined and stored for specific needs in a more sorted and customized form for reporting and analysis. In this chapter, we will discuss such a problem where a huge volume of data needs to be analyzed and stored in a more customized format for a specific downstream audience.

Problem statement

Let's assume that an e-commerce firm, ABC, wants to analyze various kinds of user interaction on its products and determine the top-selling products for each category each month. They want to provide incentives to the top-selling products in each category. They also want to provide special offers and marketing promotion tools to products with top view-to-sale ratios but are not the top-selling products. In addition, they want to market seller tools and training, as well as marketing services, to the team with the lowest-selling product in each category. Currently, ABC stores all user transactions in its transactional databases for a product, but there is no monthly data view where information about the top seller, worst seller, and the top view-to-sale ratio is available. They want an **Organized Data Layer** (ODL) to be created so that such analytical queries can easily be performed with optimum performance every month.

Analyzing the problem

Let's analyze the given problem. First, let's analyze the requirements in terms of the four dimensions of data.

First, we will try to answer the question, *what is the velocity of the data?* Here, as evident from the requirements, we need to create an ODL with monthly analyzed data. Hence, our data will be used after a month, so we have no real-time data processing requirement. So, we can safely assume that we are dealing with a batch processing problem. However, it will be helpful to know how frequently the data arrives, which will help us determine at what frequency we can schedule a batch job. So, we must ask the e-commerce firm, *how frequently will source data be provided to us?* ABC tells us that the source data will be dropped to us as CSV files on a monthly or bi-monthly basis, but never twice daily. This information is helpful to us, but that brings other questions to mind.

Now, the most obvious next question that comes to our mind is, *how big is the data/file that will be shared once or twice monthly?* Considering that each record will be an event on each transaction that any user has made on any product in the e-commerce marketplace, the data is likely to be huge. ABC tells us that transactions can be either view, cart, or purchase transactions. So, for each action, such as viewing a product, adding to the cart, and purchasing a product, there will be separate entries in the file. ABC also tells us that the number of products and categories is likely to increase in the future. From our guestimate and ABC's internal data, each file sent to us can vary from hundreds of gigabytes to terabytes of data. The data that needs to be processed is in the range of hundreds of gigabytes and terabytes, which is ideal for big data processing. Our analysis also tells that the e-commerce traffic is going to increase over time. These observations indicate that this is a big data problem. So, we need to develop a big data solution to solve this batch processing problem.

Now, we will look at the *variety of the data*. From our previous discussion, we know that data is arriving in CSV format. So, this is structured data that needs to be analyzed and processed. We will hold the discussion on the variety of data for a while as we will be taking that up during the implementation phase.

The next decision that we must make, as architects, is to choose the right platform. Should we run this application on-premise or in the cloud? There are pros and cons to both. However, there are a few vital points regarding why the cloud may be a better choice in this case:

- **Cost saving**: Running a big data job with terabytes of data will require a very good big data infrastructure to run on-premise. However, these jobs will only run once or twice a month. If we choose to create an on-premise environment, it doesn't make sense to spend so many dollars creating a clustered Hadoop infrastructure that will only be used once or twice a month, but where the infrastructure needs to be maintained and running at all times. The amount of cost and effort involved in creating and maintaining such an infrastructure doesn't justify the utilization. This will be much cheaper on the cloud, where you pay only for the resources you utilize during the job run. The cloud can give you the choice to only pay for what you use.

  For example, in the cloud, you can choose to keep your Hadoop environment running for only 2 hours daily; this way, you only pay for those 2 hours and not for the whole day. More importantly, it supports elasticity, which means you can auto-scale your number of nodes to a higher or lower number based on your usage. This gives you the flexibility to use only the required resource each time. For example, if we know that the data (which needs to be processed) will be huge in November and the jobs will take up more resources and time in November, we can increase the resource capacity for November and bring it down when the volume reduces to a normal level. Such capabilities of cloud technologies enable huge cost savings on the overall execution of the system (especially **capital expenditure (CapEx)** costs).

- **Seasonal variation in workloads**: Usually, in an e-commerce site, the activity during the holiday season or festival season is high, while at other times, the activity is low. User activity directly impacts the size of the file for that month. So, we must be able to scale the infrastructure up and down as we need. This can easily be achieved in the cloud.

- **Future elasticity**: As one of the requirements clearly states that the number of products and categories is likely to increase, this means we will need to scale up both processing and storage capacities in the future. While such changes require a lot of time and resources in on-premise environments, this can easily be achieved in the cloud.

- **Lack of sensitive data**: There is no specific federal or **Protected Health Information** (PHI) data involved in our use case that needs to be encrypted or tokenized before it is stored in the cloud. So, we should be good with legal and data security requirements.

Although we can choose any public cloud platform, for our convenience, we will use AWS as our cloud platform in this book.

Architecting the solution

Now that we have gathered and analyzed our requirements, let's try building the architecture for the solution. To architect this solution, we need to answer the following questions:

- Where should we store or land the input data?
- How should we process the input data?
- Where and how should we store the output data/ODL?
- How should we provide a querying interface to the ODL?
- How and when should we schedule a processing job?

Let's see what options we have for storing the input data. We can store the data in one of the following services:

- **S3**: **S3** or **Simple Storage Service** is a very popular object storage service. It is also cheap and very reliable.
- **EMR/EC2 attached EBS volumes**: **Elastic Block Storage** (EBS) is a block storage solution where storage volumes can be attached to any virtual server, such as an EC2 instance. For a big data solution, if you use **Elastic Map Reduce** (EMR), EBS volumes can be attached to each participating EC2 node in that EMR cluster.
- **Elastic File System** (EFS): EFS is a shared filesystem that's often attached to a NAS server. It is usually used for content repositories, media stores, or user home directories.

Let's discuss the different factors to consider before choosing your storage.

Factors that affect your choice of storage

Cost is an important factor that we need to consider when choosing any cloud component. However, let's look at factors other than cost that affect our choice of storage. These factors are as follows:

- **Performance**: Both EBS and EFS can perform faster than S3 in terms of IOPS. Although performance is slower in S3, it's not significantly slower to read the data from other storage options. From a performance perspective, an EFS or EBS volume will still be preferred.

- **Scalability**: Although all three storage options are scalable, S3 has the most seamless scalability without any manual effort or interruption. Since scalability is one of our important needs as our data grows over time and there is a possibility of bigger file sizes in the future (according to the requirements), from this perspective, S3 is a clear winner.

- **Life cycle management**: Both S3 and EFS have life cycle management features. Suppose you believe that older files (older than a year) need to be archived; these programs can seamlessly move to another cheaper storage class, which provides seamless archival storage as well as cost savings.

- **Serverless architecture support**: Both S3 and EFS provide serverless architecture support.

- **High availability and robustness**: Again, both S3 and EFS are highly robust and available options for storage. In this regard, EBS is not on par with the other two storage options.

- **Big data analytic tool compatibility**: Reading and writing data from an S3 or EBS volume is much easier from big data processing engines such as Spark and MapReduce. Also, creating external Hive or Athena tables is much easier if the data resides in S3 or EBS.

As we can see, both S3 and EFS seem to be promising options to use. Now, let's look at how crucial cost is in determining cloud storage solutions.

Determining storage based on cost

One of the most important tools for any cloud solution architect is a cost estimator or pricing calculator. As we are using AWS, we will use AWS Pricing Calculator: `https://calculator.aws/#/`.

We will use this tool to compare the cost of storing input data in EFS versus S3 storage. In our use case, we'll assume that we get 2 TB of data per month and that we must store monthly data for 3 months before we can archive it. We also need to store data for up to 1 year. Let's see how our cost varies based on our choice of storage.

Here, for either kind of storage, we will use **S3 Intelligent-Tiering** (which supports automatic life cycle management and reduces cost) to do the calculation. It asks for the average data storage per month and the amount stored in the frequent access layer, infrequent access layer, and archive layers.

To calculate the average data storage required per month, 2 TB of new data per month gets generated for our use case. So, we have 2 TB of data to store in the first month, 4 TB of data to store in the second, 6 TB of data to store in the third, and so on. So, to calculate the average data, we must add all the storage requirements for each month together and divide the result by 12. The mathematical equation for this is as follows:

$$DataStorage_{avg} = \frac{2 * \sum_{i=1}^{12} i}{12}$$

The preceding formula gives us a 13 TB per month calculation. Now, it asks us for the percentage stored in the frequent access layer – the layer that we will read the data from. The data in the frequent access layer can only be 2 TB for each month (which is around 15% of 13 TB). Using these values, we can calculate the estimated cost, as shown in the following screenshot:

Figure 5.1 – AWS S3 cost estimation tool

Using the previously mentioned calculation, ballpark estimates for Amazon S3 come to 97.48 USD per month on average. However, a similar calculation for Amazon EFS would cost 881.92 USD. This depicts that using EFS will be nine times costlier than using Amazon S3.

So, looking at the cost, combined with other parameters, we can safely decide on choosing Amazon S3 as our input storage. Based on a similar set of logic and calculations, we can store the ODL in S3 as well.

However, a discussion about the storage layer for the output data is incomplete without deciding on the schema and format of the output files. Based on the requirements, we can conclude that the output data should have the following columns:

- `year`
- `month`
- `category_id`
- `product_id`
- `tot_sales`

- `tot_onlyview`
- `sales_rev`
- `rank_by_revenue`
- `rank_by_sales`

It is advisable to partition the table on a yearly and monthly basis since most of the queries on the ODL will be monthly. Now that we have finalized all the details of the storage layer, let's discuss the processing layer of the solution.

Now, we must look at the options that are available in AWS for processing a big data batch job. Primarily, there are two native alternatives. One is running Spark on EMR clusters, while the other is running Glue jobs (Glue is a fully managed serverless AWS service). AWS Glue allows you to write a script in Scala or Python and trigger Glue jobs either through the AWS Management Console or programmatically. Since we are interested in implementing the solution in Java, AWS Glue is not an option for us. Also, AWS Glue scripts have less portability and a higher learning curve. Here, we will stick to Spark on EMR. However, Spark jobs can be run on an EMR cluster in two ways:

- The classical approach of running `spark submit` from the command line
- The cloud-specific approach of adding a `spark submit` step

Now, let's see how the cost matrix helps us determine which approach we should take to submit a Spark job among the two options mentioned earlier.

The cost factor in the processing layer

The first option is to keep the EMR cluster up and running all the time and trigger the `spark submit` command using a shell script from a cronjob trigger at specific time intervals (very similar to what we would do on an on-premise Hadoop cluster). Such a cluster is known as a *persistent EMR cluster*.

The second option is to add an EMR step to run the Spark job while creating the cluster and then terminate it once it has run successfully. This kind of EMR cluster is known as a *transient EMR cluster*. Let's see how the cost estimates vary for each option.

In an EMR cluster, there are three types of nodes: *master node*, *core node*, and *task node*. The master node manages the cluster and acts as the NameNode and the Jobtracker. The core node acts as the DataNode, as well as the worker node, which is responsible for processing the data. TaskNodes are optional, but they are required for separate task tracker activities.

Due to their nature of work, usually, selecting a compute-optimized instance works great for the master node, while a mixed instance works best for the core nodes. In our calculation, we will use the c4.2xlarge instance type for the master node and the m4.4xlarge instance type for the core nodes. If we need four core nodes, a persistent EMR cluster would cost us around 780 USD per month. A similar configuration on a transient EMR cluster would only cost around 7 USD per month, considering the job runs two or three times a month with a job run duration not exceeding 2 hours each. As we can see, the second option is nearly 100 times more cost-effective. Therefore, we will choose a transient EMR cluster.

Now, let's figure out how to create and schedule the transient EMR clusters. In our use case, the data arrives in the S3 buckets. Each successful creation event in an S3 bucket generates an event that can trigger an AWS Lambda function. We can use such a Lambda function to create the transient cluster every time a new file lands in the landing zone of the S3 bucket.

Based on the preceding discussion, the following diagram depicts the architecture of the solution that's been proposed:

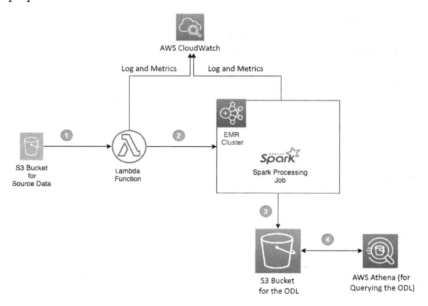

Figure 5.2 – Solution architecture

As shown in the preceding diagram, the input data lands in the S3 bucket (also known as landing zone) as the source data. Here, one source data file arrives twice a month. The architecture diagram depicts four steps denoted by numerals. Let's look at these steps in more detail:

1. A CloudWatch event is generated when an incoming source file is completely written in the S3 bucket. This generates a Lambda trigger, which, in turn, invokes a Lambda function.

2. The Lambda function receives the creation event records and creates a transient EMR cluster with a step configured to run a Spark job to read and process the new input file(s).

3. The Spark job in the EMR step reads and processes the data. Then, it writes the transformed output data to the S3 bucket for the ODL layer. Upon successfully terminating the Spark step, the transient cluster gets terminated.

4. All the data residing in the ODL layer will be exposed as Athena tables that can be queried for any analytical purposes.

This gives us a very simple yet powerful architecture to solve such big data batch processing problems. The logs and metrics in all the processing and storage components will be captured by AWS CloudWatch logs. We can further improve this architecture by adding auditing and alert features using CloudWatch logs and metrics.

> **Important note**
>
> It is advisable to use the Parquet format as the output storage format because of the following factors:
>
> **Cost saving and performance**: Since multiple output columns can potentially have low cardinality, the ODL data storage format should be a columnar format, which can give cost savings as well as better performance.
>
> **Technology compatibility**: Since we are dealing with big data processing, and our processing layer is Spark-based, Parquet will be the most suitable data format to use for the ODL layer.

Now that we have analyzed the problem and developed a robust, reliable, and cost-effective architecture, let's implement the solution.

Implementing the solution

The first step of any implementation is always understanding the source data. This is because all our low-level transformation and cleansing will be dependent on the variety of the data. In the previous chapter, we used DataCleaner to profile the data. However, this time, we are dealing with big data and the cloud. DataCleaner may not be a very effective tool for profiling the data if its size runs into the terabytes. For our scenario, we will use an AWS cloud-based data profiling tool called AWS Glue DataBrew.

Profiling the source data

In this section, we will learn how to do data profiling and analysis to understand the incoming data (you can find the sample file for this on GitHub at `https://github.com/PacktPublishing/Scalable-Data-Architecture-with-Java/tree/main/Chapter05`. Follow these steps:

1. Create an S3 bucket called `scalabledataarch` using the AWS Management Console and upload the sample input data to the S3 bucket:

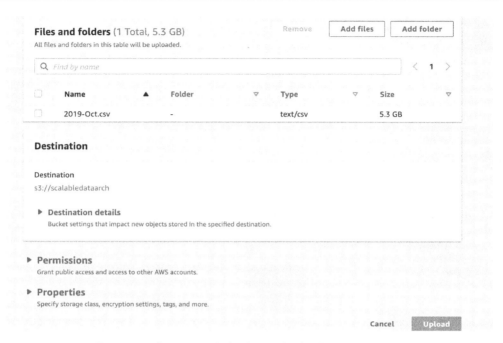

Figure 5.3 – Creating an S3 bucket and uploading the input file

2. From the AWS Management Console, go to the AWS Glue DataBrew service. Click on the **DATASET** side tab. Then, click the **Connect new dataset** button. A dialog box similar to the one shown in the following screenshot will appear. Select **Amazon S3** and then enter the source data path of the S3 bucket. Finally, click the **Create Dataset** button to create a new dataset:

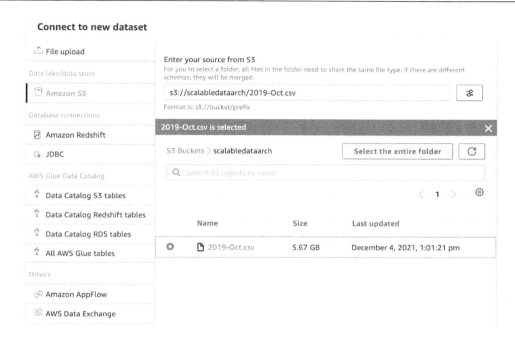

Figure 5.4 – Adding a new dataset to AWS Glue DataBrew

3. Now, let's create a data profiling job using the dataset that we've added. First, select the newly added dataset and go to the **Data profile overview** tab, as shown in the following screenshot:

Figure 5.5 – The Data profile overview tab

Now, click on the **Run data profile** button, which will take you to a **Create job** popup. Enter the job name and choose to run the sample using the **Full dataset** option, as shown here:

Figure 5.6 – Creating a data profiling job

Set the output location as `scalablearch-dataprof` (our S3 bucket). The output files of the data profiling job will be stored here:

Figure 5.7 – Configuring the data profiling job

Then, configure **Dataset level configurations**, as shown in the following screenshot. Since we want to find the correlation between `product_id`, `category_id`, and `brand`, we have configured them accordingly, as shown in the following screenshot:

Dataset level configurations
Choose the dataset-level statistics that you want to include in your generated data profile.
Default
Duplicate values
Enabled
Advanced statistics
PII statistics
Disabled
Correlations widget
Enabled, custom columns
Correlations widget
High impact
Enable correlations matrix
Matrix visualization of correlation coefficient (r). Correlation coefficient (r) defines how closely two variables are related. It ranges from -1.
relationship between the variables.
Correlation algorithm
Algorithm used to identify correlations
Pearsons correlation
Selected columns
Select the columns to calculate correlations for
All columns
First
10
numeric columns
Custom columns
Column name
RegEx match
Enter column name
Add
You can add a maximum of 25 column names or regular expressions
Name: product_id
Name: category_id
Name: brand

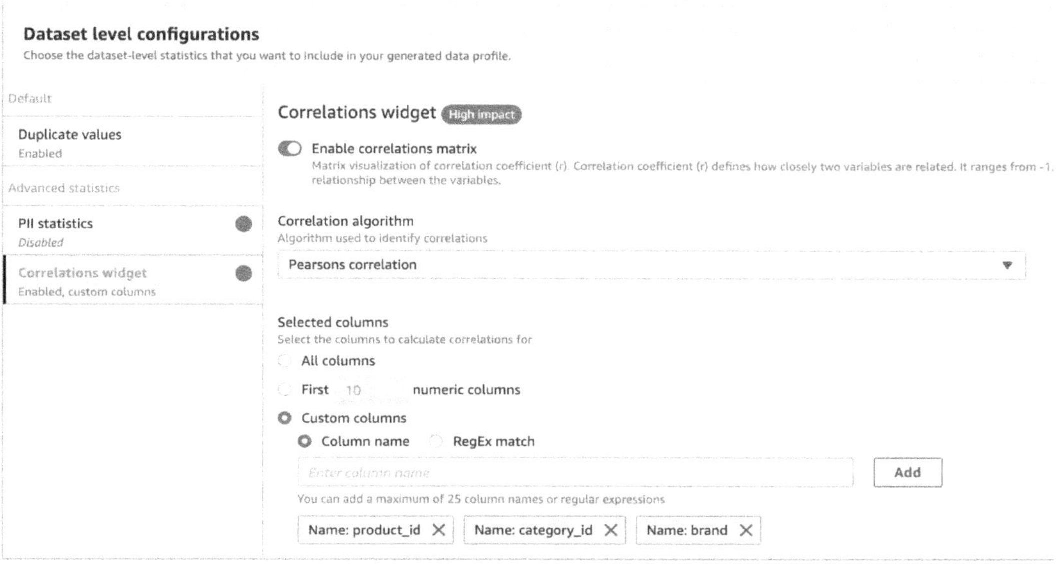

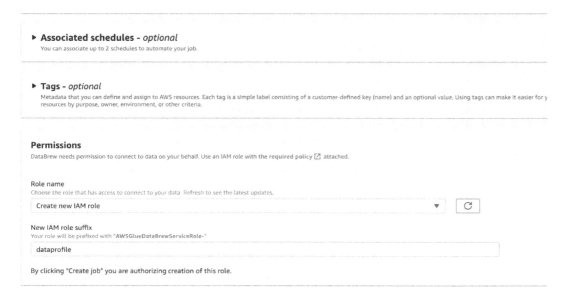

Figure 5.8 – Configuring the Correlations widget

Then, we must set up the security roles for the data profiling job. Once you've done this, click the **Create Job** button to create the data profiling job:

Associated schedules - optional
You can associate up to 2 schedules to automate your job.
Tags - optional
Metadata that you can define and assign to AWS resources. Each tag is a simple label consisting of a customer-defined key (name) and an optional value. Using tags can make it easier for y
resources by purpose, owner, environment, or other criteria.
Permissions
DataBrew needs permission to connect to data on your behalf. Use an IAM role with the required policy attached.
Role name
Choose the role that has access to connect to your data. Refresh to see the latest updates.
Create new IAM role
New IAM role suffix
Your role will be prefixed with "AWSGlueDataBrewServiceRole-"
dataprofile
By clicking "Create job" you are authorizing creation of this role.

Figure 5.9 – Setting the security permissions for the data profiling job

4.   Finally, we can see the newly created data profiling job in the **Profile jobs** tab. We can run the data profiling job by clicking the **Run job** button on this screen:

Figure 5.10 – Data profiling job created and listed

Once the job successfully runs, we can go to the dataset, open the **Data lineage** tab, and view the data lineage, as well as the time interval before which the last successful data profiling job ran:

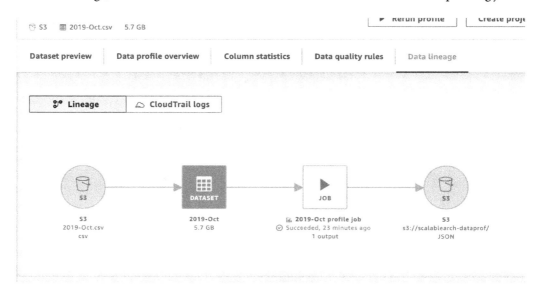

Figure 5.11 – Lineage of the data profiling job

5.   We can visualize the report to find missing values, cardinalities, the correlation between columns, and the distribution of the data. These metrics help us determine whether there are anomalies in the data that need to be cleaned up or if there are missing values and if they need to be handled. It also helps us understand the quality of the data that we are dealing with. This helps us do proper cleansing and transformation so that it doesn't give us surprises later during our implementation. The following screenshot shows some sample metrics that AWS Glue DataBrew displays:

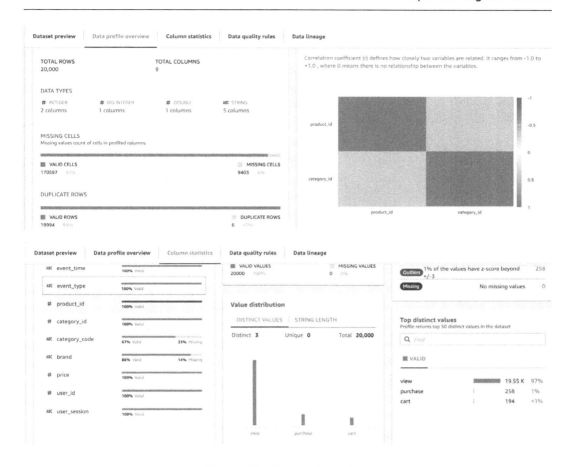

Figure 5.12 – Data profile metrics

Here, we can see some useful statistics. For example, `event_type` has no noise and it has a very low cardinality. It also shows that the data is not uniformly distributed by this column.

Now that we have analyzed the data, let's develop the Spark application that will process the records.

Writing the Spark application

Based on the analysis in the previous section, we will create the incoming record schema string. Then, we will use that schema string to read the incoming data, as shown in the following code snippet:

```
private static final String EVENT_SCHEMA = "event_time
TIMESTAMP,event_type STRING,product_id LONG,category_id
LONG,category_code STRING,brand STRING,price DOUBLE,user_id
LONG,user_session STRING";
```

```
. . .
Dataset<Row> ecommerceEventDf = spark.read().
option("header","true").schema(EVENT_SCHEMA).csv(inputPath);
```

Then, we will calculate the total sales and total views for each product using the count_if aggregation function of Spark, as shown in the following code snippet:

```
Dataset<Row> countAggDf = spark.sql("select year(event_time)
as year,month(event_time) as month,category_id,product_
id,count_if(event_type='purchase') as tot_sales,count_if(event_
type='view') as tot_onlyview from ecommerceEventDf where event_
type!='cart' group by year,month,category_id,product_id ");
```

We will create another DataFrame to calculate the total revenue for only the purchase events. The following code snippet shows how to do this:

```
Dataset<Row> revenueAggDf = spark.sql("select year(event_
time) as year,month(event_time) as month,category_id,product_
id,sum(price) as sales_rev from ecommerceEventDf where event_
type='purchase' group by year,month,category_id,product_id");
```

Now, we will combine the countAggDf and revenueAggDf DataFrames using a LEFT OUTER JOIN SparkSQL query, as shown in the following code snippet. The null values for total_sales for the product that didn't have a single sale are set to 0.0 using the na.fill() method of Spark:

```
Dataset<Row> combinedAggDf = spark.sql("select cadf.year,cadf.
month,cadf.category_id,cadf.category_id,cadf.product_id,tot_
sales,tot_onlyview,sales_rev from countAggDf cadf LEFT OUTER
JOIN revenueAggDf radf ON cadf.year==radf.year AND cadf.month==
radf.month AND cadf.category_id== radf.category_id AND cadf.
product_id == radf.product_id");

Dataset<Row> combinedEnrichAggDf = combinedAggDf.na().
fill(0.0,new String[]{"sales_rev"});
```

Now, we will apply window functions on the resultant combinedEnrichedDf DataFrame to derive the columns – that is, rank_by_revenue and rank_by_sales:

```
Dataset<Row> finalTransformedDf = spark.sql("select
year,month,category_id,product_id,tot_sales,tot_onlyview,sales_
rev,dense_rank() over (PARTITION BY category_id ORDER BY sales_
rev DESC) as rank_by_revenue,dense_rank() over (PARTITION BY
category_id ORDER BY tot_sales DESC) as rank_by_sales from
combinedEnrichAggDf");
```

The result is ready and is in the same format as the output. So, we must write the transformed data to the output S3 bucket using Parquet format while ensuring it's partitioned by year and month:

```
finalTransformedDf.write().mode(SaveMode.Append).
partitionBy("year","month").parquet(outputDirectory);
```

The full source code for this application is available on GitHub at `https://github.com/PacktPublishing/Scalable-Data-Architecture-with-Java/tree/main/Chapter05/sourcecode/EcommerceAnalysis`.

In the next section, we will learn how to deploy and run the Spark application on an EMR cluster.

Deploying and running the Spark application

Now that we have developed the Spark job, let's try to run it using a transient EMR cluster. First, we will create an EMR cluster manually and run the job. To create the transient EMR cluster manually, follow these steps:

1. First, build the Spark application JAR file and upload it to an S3 bucket.
2. Go to the AWS Management Console for AWS EMR. Click the **Create Cluster** button to create a new transient cluster manually.
3. Set up the EMR configuration. Make sure that you set **Launch mode** to **Step execution**. Make sure that you select **emr-6.4.0** as the value of the **Release** field in the **Software configuration** section. Also, for **Add Steps**, choose **Spark application** for **Step type**. Leave all the other fields as-is. Your configuration should look as follows:

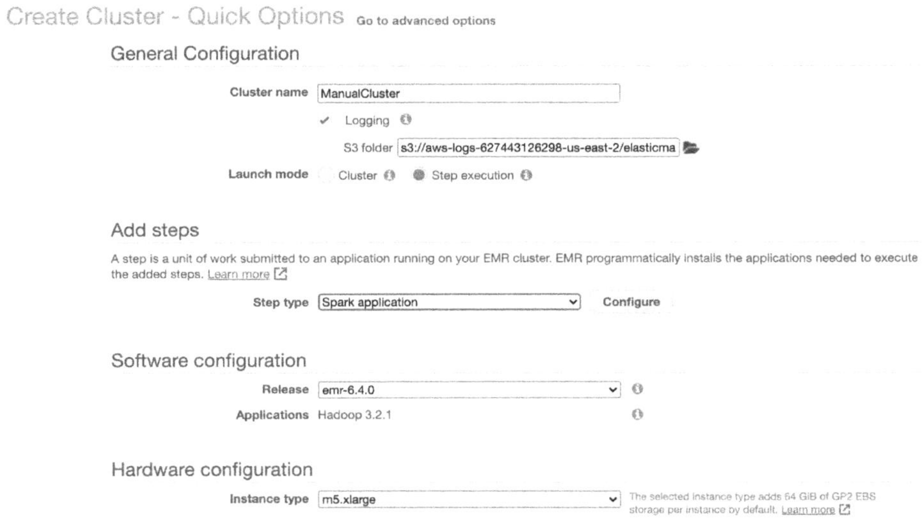

Figure 5.13 – Manual transient EMR cluster creation

4. Now, to add the Spark step, click on the **Configure** button. This will make a dialog box appear where you can enter various Spark step-related configurations, as shown in the following screenshot:

Figure 5.14 – Adding a Spark step to the EMR cluster

Please ensure that you specify the driver class name in the **Spark-submit options** area and provide the necessary information in the **Application location*** and **Arguments** boxes.

5. Click **Add** to add the step. Once added, it will look similar to what's shown in the following screenshot. Then, click **Create Cluster**. This will create the transient cluster, run the Spark job, and terminate the cluster once the Spark job has finished executing:

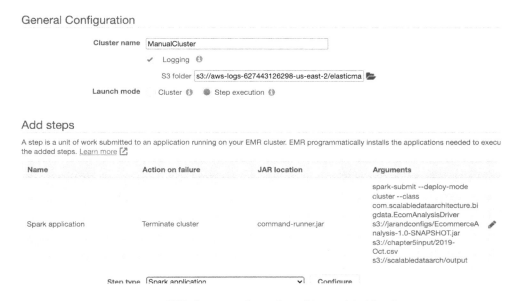

Figure 5.15 – EMR cluster configuration with an added Spark step

6.   Once it has successfully run, you will see that the job succeeded in the **Steps** tab of the cluster:

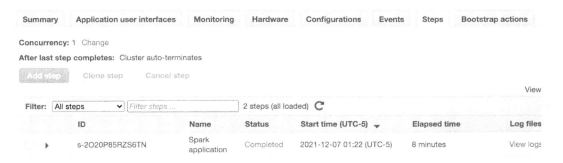

Figure 5.16 – Job monitoring in the EMR cluster

> **Troubleshooting Spark errors**
>
> A Spark job runs on huge volumes of data and can throw multiple exceptions. It can also report multiple stage failures, such as `OutOfMemoryException`, large frame errors, throttling errors from multipart files uploaded in AWS S3, and so on. Covering all of these is beyond the scope of this book. However, you can refer to a very concise Spark troubleshooting guide at `https://docs.qubole.com/en/latest/troubleshooting-guide/spark-ts/troubleshoot-spark.html#troubleshooting-spark-issues` for more information.

Now that we have deployed and run the Spark application manually, let's automate how Spark jobs are created and run by implementing a Lambda trigger.

Developing and testing a Lambda trigger

AWS Lambda functions are fully managed serverless services that help process information. They are supported by multiple languages such as Python, JavaScript, Java, and so on. Although Python or JavaScript runtimes are faster, we will use the Java runtime in this book to implement the solution (since we are focusing on Java-based implementations in this book).

To write a Lambda function that will react to an S3 event, we must create a Java class that implements the `RequestHandler` interface and takes `S3Event` as its generic `Input` type, as shown in the following code block:

```
import com.amazonaws.services.lambda.runtime.RequestHandler;
import com.amazonaws.services.lambda.runtime.events.S3Event;
. . .



public class Handler implements RequestHandler<S3Event,
Integer> {
. . .
```

In this class, we must implement the `handleRequest` method of the `RequestHandler` interface. In the `handleRequest` method, we loop through each `S3EventNotificationRecord`, which denotes a new file being created or updated. We collect all the S3 object names attached to this `S3EventNotificationRecord` in `S3ObjectNames`. For each distinct S3 object name present in `S3ObjectNames`, we create and launch an AWS transient EMR cluster. The following code snippet shows the implementation of the `handleRequest` method:

```
Set<String> s3ObjectNames = new HashSet<>();
for (S3EventNotificationRecord record:
        s3Event.getRecords()) {
    String s3Key = record.getS3().getObject().getKey();
    String s3Bucket = record.getS3().getBucket().getName();
    s3ObjectNames.add("s3://"+s3Bucket+"/"+s3Key);
}

s3ObjectNames.forEach(inputS3path ->{
    createClusterAndRunJob(inputS3path,logger);
```

```
});
```

Now, let's look at the implementation of the `createClusterAndRunJob` method. This takes two arguments: `inputS3path` and the Lambda logger. This method uses AWS SDK to create an `ElasticMapReduce` object. This method uses the `StepConfig` API to build a `spark submit` step. Then, it uses all the configuration details, along with `SparkSubmitStep`, to configure the `RunJobFlowRequest` object.

Finally, we can submit a request to create and run an EMR cluster using the `runJobFlow` method of the `ElasticMapReduce` object, as shown here:

```
private void createClusterAndRunJob(String inputS3path,
LambdaLogger logger) {
    //Create a EMR object using AWS SDK
    AmazonElasticMapReduce emr =
AmazonElasticMapReduceClientBuilder.standard()
            .withRegion("us-east-2")
            .build();


    // create a step to submit spark Job in the EMR cluster to
be used by runJobflow request object
    StepFactory stepFactory = new StepFactory();


    HadoopJarStepConfig sparkStepConf = new
HadoopJarStepConfig()
            .withJar("command-runner.jar")
            .withArgs("spark-submit","--deploy-
mode","cluster","--class","com.scalabledataarchitecture.
bigdata.EcomAnalysisDriver","s3://jarandconfigs/
EcommerceAnalysis-1.0-SNAPSHOT.jar",inputS3path,"s3://
scalabledataarch/output");


    StepConfig sparksubmitStep = new StepConfig()
            .withName("Spark Step")
            .withActionOnFailure("TERMINATE_CLUSTER")
            .withHadoopJarStep(sparkStepConf);


    //Create an application object to be used by runJobflow
request object
```

```java
Application spark = new Application().withName("Spark");

//Create a runjobflow request object
RunJobFlowRequest request = new RunJobFlowRequest()
        .withName("chap5_test_auto")
        .withReleaseLabel("emr-6.4.0")
        .withSteps(sparksubmitStep)
        .withApplications(spark)
        .withLogUri(...)
        .withServiceRole("EMR_DefaultRole")
        ...                     ;

//Create and run a new cluster using runJobFlow method
RunJobFlowResult result = emr.runJobFlow(request);
logger.log("The cluster ID is " + result.toString());
}
```

Now that we developed the Lambda function, let's deploy, run, and test it:

1. Create an IAM security role for the Lambda function to trigger the EMR cluster, as shown in the following screenshot:

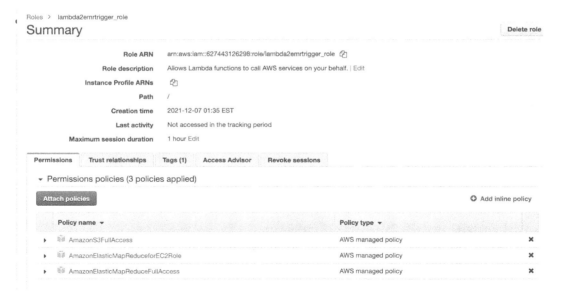

Figure 5.17 – Creating a new IAM role

2. Create a Lambda function using the AWS Management Console. Please provide the name and runtime of the function, as shown in the following screenshot:

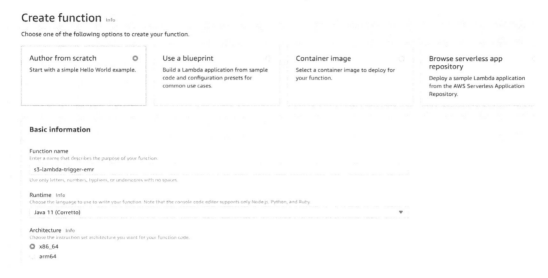

Figure 5.18 – Creating an AWS Lambda function

3. While creating the Lambda function, please make sure that you change the default execution role to the IAM role you created in *Step 1*:

Figure 5.19 – Setting the IAM role to an AWS Lambda function

4. Now, you must add an S3 trigger for the Lambda function, as shown in the following screenshot. Make sure that you enter the proper bucket name and prefix where you will push your source files:

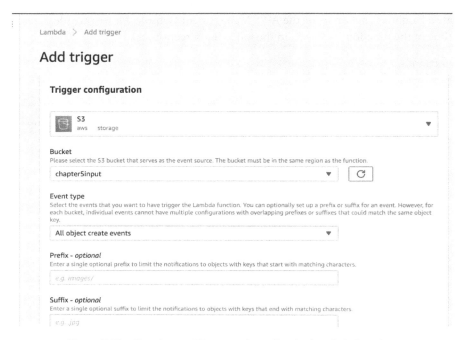

Figure 5.20 – Creating an S3 event trigger for the Lambda function

5.   Then, you must build the JAR file locally from the Lambda function we developed using our Maven Java project (the full source code for the project can be found at `https://github.com/PacktPublishing/Scalable-Data-Architecture-with-Java/tree/main/Chapter05/sourcecode/S3lambdaTriggerEmr`). Once the JAR file has been built, you must upload it using the **Upload from | .zip or .jar file** option, as shown in the following screenshot:

Figure 5.21 – Deploying an AWS Lambda JAR file

6. Now, you can test the whole workflow by placing a new data file in the S3 bucket mentioned in the S3 trigger. Once the Lambda function executes, it creates a transient EMR cluster where the Spark job will run. You can monitor the metrics of the Lambda function from the AWS Management Console:

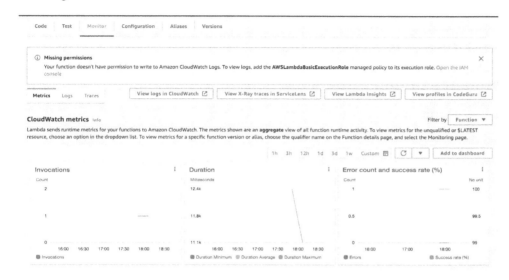

Figure 5.22 – Monitoring the AWS Lambda function

You can monitor the Spark application from the AWS EMR management console by looking through the **Persistent user interfaces** options in the transient cluster's **Summary** tab, as shown in the following screenshot:

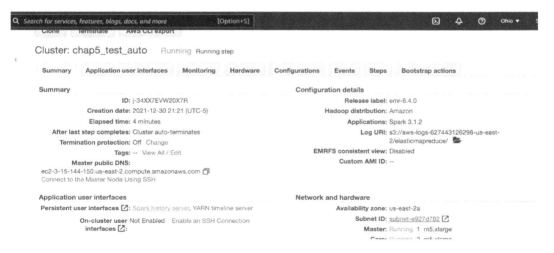

Figure 5.23 – EMR Cluster management console

> **Troubleshooting a Lambda function**
>
> In the real world, you may have trouble when invoking or executing a Lambda function. A very concise guide to troubleshoot all such issues has been published by AWS. For more information, check out `https://docs.aws.amazon.com/lambda/latest/dg/lambda-troubleshooting.html`.

Now, let's see if we can further optimize the Spark application by monitoring the Spark job.

Performance tuning a Spark job

We can investigate the Spark UI to see its **directed acyclic graph (DAG)**. In our case, our DAG looks like this:

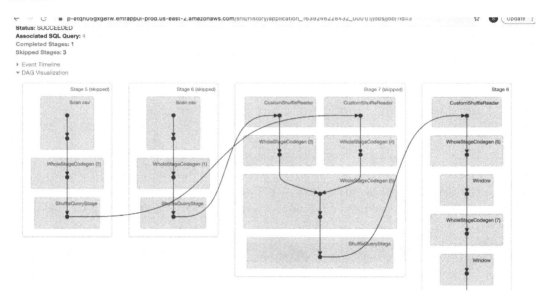

Figure 5.24 – DAG of the Spark job

As we can see, both *Stage 5* and *Stage 6* are performing the same job of scanning and reading the CSV into a DataFrame. This is because we have a DataFrame called `ecommerceEventDf` that is being used to derive two different DataFrames. Both derived DataFrames calculate `ecommerceEventDf` separately due to Spark's lazy evaluation technique, which causes the performance to slow down. We can overcome this issue by persisting the `ecommerceEventDf` DataFrame, as shown in the following code snippet:

```
ecommerceEventDf.persist(StorageLevel.MEMORY_AND_DISK());
```

After making this change, the new DAG will look as follows:

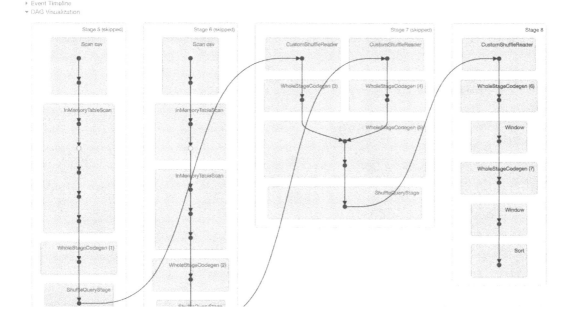

Figure 5.25 – Optimized DAG of the Spark job

In the new DAG, there's a green dot in the `InMemoryTableScan` task. This green dot represents the in-memory persistence of the data by Spark so that it doesn't scan the CSV file twice, thus saving processing time. In this use case, it will speed up the performance of the Spark job by around 20%.

Now that we have implemented and tested our solution, let's learn how to build an Athena table on top of the output folder and enable easy querying of the results.

Querying the ODL using AWS Athena

In this section, we will learn how to perform data querying on the ODL that we have created using our architecture. We will focus on how to set up Athena on our output folder to do easy data discovery and querying:

1. Navigate to AWS Athena via the AWS Management Console. Click on **Explore the query editor**. First, go to the **Manage settings** form of the **Query editor** area and set up an S3 bucket where the query results can be stored. You can create an empty bucket for this purpose:

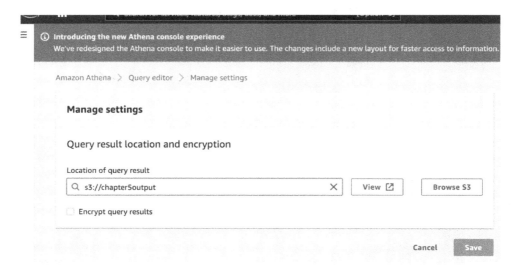

Figure 5.26 – Setting up AWS Athena

2. We will create an Athena table on top of our S3 output bucket. For this, we will create a DDL to create a table called ecom_odl, which is a partitioned table on the year and month columns. The DDL of the table can be seen in the following code snippet:

```
CREATE EXTERNAL TABLE IF NOT EXISTS ecom_odl(
      category_id bigint,
      product_id bigint,
      tot_sales bigint,
      tot_onlyview bigint,
      sales_rev double,
      rank_by_revenue int,
      rank_by_sales int
) PARTITIONED BY (year int, month int) STORED AS parquet
LOCATION 's3://scalabledataarch/output/';
```

We will run this DDL statement in the **Query editor** area of Athena to create the table shown in the following screenshot:

Figure 5.27 – Creating an Athena table based on the output data

3. Once the table has been created, we need to add the partition. We can do this by using the MSCK REPAIR command (similar to Hive):

    MSCK REPAIR TABLE ecom_odl

4. Upon running the previous command, all partitions are auto-discovered from the S3 bucket. Now, you can run any query on the ecom_odl table and get the result. As shown in the following screenshot, we run a sample query to find the top three products by revenue for each category in October 2019:

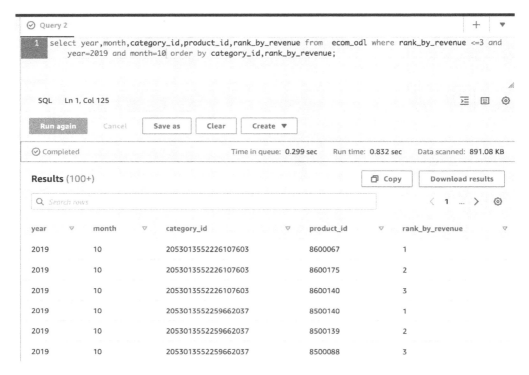

Figure 5.28 – Querying ODL using an Athena table

With that, we have successfully architected, designed, and developed a big data batch processing solution and created an interface for the downstream teams to query our analyzed data using AWS Athena. Now, let's summarize what we have learned in this chapter.

Summary

In this chapter, we learned how to analyze a problem and identified that it was a big data problem. We also learned how to choose a platform and technology that will be performance-savvy, optimized, and cost-effective. We learned how to use all these factors judiciously to develop a big data batch processing solution in the cloud. Then, we learned how to analyze, profile, and draw inferences from big data files using AWS Glue DataBrew. After that, we learned how to develop, deploy, and run a Spark Java application in the AWS cloud to process a huge volume of data and store it in an ODL. We also discussed how to write an AWS Lambda trigger function in Java to automate the Spark jobs. Finally, we learned how to expose the processed ODL data through an AWS Athena table so that downstream systems can easily query and use the ODL data.

Now that we have learned how to develop optimized and cost-effective batch-based data processing solutions for different kinds of data volumes and needs, in the next chapter, we will learn how to effectively build solutions that help us process and store data in real time.

6

Architecting a Real-Time Processing Pipeline

In the previous chapter, we learned how to architect a big data solution for a high-volume batch-based data engineering problem. Then, we learned how big data can be profiled using Glue DataBrew. Finally, we learned how to logically choose between various technologies to build a Spark-based complete big data solution in the cloud.

In this chapter, we will discuss how to analyze, design, and implement a real-time data analytics solution to solve a business problem. We will learn how the reliability and speed of processing can be achieved with the help of distributed messaging systems such as Apache Kafka to stream and process the data. Here, we will discuss how to write a Kafka Streams application to process and analyze streamed data and store the results of a real-time processing engine in a NoSQL database such as MongoDB, DynamoDB, or DocumentDB using Kafka connectors.

By the end of this chapter, you will know how to build a real-time streaming solution to predict the risk category of a loan application using Java and Kafka-related technologies. You will also know how a real-time data analytics problem is designed and architected. Throughout this journey, you will learn how to publish events to Kafka, analyze that data using Kafka Streams, and store the result of the analytics in MongoDB in real time. By doing so, you will know how to approach a real-time data engineering problem and build an effective streaming solution.

In this chapter, we're going to cover the following main topics:

- Understanding and analyzing the streaming problem
- Architecting the solution
- Implementing and verifying the design

Technical requirements

To complete this chapter, you'll need the following:

- Prior knowledge of Java
- Java 1.8 or above, Maven, Apache Kafka, and PostgreSQL installed on your local system
- A MongoDB Atlas subscription in the cloud
- IntelliJ IDEA Community or Ultimate edition installed on your local system

Understanding and analyzing the streaming problem

So far, we have looked at data engineering problems that involve ingesting, storing, or analyzing the stored data. However, in today's competitive business world, online web apps and mobile applications have made consumers more demanding and less patient. As a result, businesses must adapt and make decisions in real time. We will be trying to solve such a real-time decision-making problem in this chapter.

Problem statement

A financial firm, XYZ, that offers credit cards, has a credit card application that works in real time and uses various user interfaces such as mobile and online web applications. Since customers have multiple options and are less patient, XYZ wants to make sure that the credit loan officer can decide on credit card approval in a split second or in real time. To do that, the application needs to be analyzed and a credit risk score needs to be generated for each application. This risk score, along with the necessary application parameters, will help the credit loan officer decide quickly.

Analyzing the problem

Let's analyze the given problem. First, let's analyze the requirements in terms of the four dimensions of data.

First, we will try to answer the question, *what is the velocity of the data?* This is the most important factor of this problem. As evident from the problem statement, unlike our previous problems, source data is being received in real time and the data analysis also needs to happen in real time. This kind of problem is well suited for a real-time streaming solution.

Now, the next dimension that we need to discuss is *volume*. However, since our problem involves streaming data, it doesn't make sense to discuss the total volume of the data. Rather, we should be answering questions such as, *how many applications are submitted every minute or every hour on average, as well as at peak times? Will this volume increase in the future? If it does, how many times and how frequently it is likely to increase?* We should go back to the client with these questions. Often, in a business, these answers are not readily available if the client is creating a real-time pipeline for the first time. In such a scenario, we should ask for the most granular average data velocity information (in this case, the number of applications filed) available with the client – for each day, week, or month and then calculate the average expected volume in a minute. Also, to understand the increase in volume, we can ask about the target projections as far as sales are concerned over a year and try to predict the volume increase.

Let's suppose that the client is getting one million applications per day and that their target is to increase sales by 50% over the next 2 years. Considering that the usual approval rate is 50%, we can expect a two times increase in the application submission rate. This would mean that we could expect a volume of 2 million applications per day in the future.

Since our solution needs to be real-time, must process more than a million records, and the volume is likely to increase in the future, the following characteristics are essential for our streaming solution:

- Should be robust
- Should support asynchronous communication between various systems within the solution and external source/sink
- Should ensure zero data loss
- Should be fault tolerant as we are processing data in real time
- Should be scalable
- Should give great performance, even if the volume increases

Keeping all these factors in mind, we should choose a pub-sub messaging system as this can ensure scalability, fault tolerance, higher parallelism, and message delivery guarantees. Distributed messaging/streaming platforms such as Apache Kafka, AWS Kinesis, and Apache Pulsar are best suited to solve our problem.

Next, we will be focusing on the *variety* of the data. In a typical streaming platform, we receive data as events. Each event generally contains one record, though sometimes, it may contain multiple records. Usually, these events are transmitted in platform-independent data formats such as JSON and Avro. In our use case, we will receive the data in JSON format. In an actual production scenario, there's a chance that the data may be in Avro format.

One of the challenges that real-time streaming solutions face is the *veracity* of the data. Often, veracity is determined based on the various possibilities of noise that can come from the data. However, accurate analysis of the veracity happens as a real-time project gets implemented and tests are run with real data. As with many software engineering solutions, real-time data engineering solutions mature over time to handle noise and exceptions.

For the sake of simplicity, we will assume that our data that is getting published in the input topic is already clean, so we won't discuss veracity in our current use case. However, in the real world, the data that is received over the input topic contains anomalies and noise, which needs to be taken care of. In such cases, we can write a Kafka Streams application to clean and format the data and put it in a processing topic. Also, erroneous records are moved to the error topic from the input topic; they are not sent to the processing topic. Then, the streaming app for data analytics consumes the data from the processing topic (which contains clean data only).

Now that we have analyzed the dimensions of data for this problem and have concluded that we need to build a real-time streaming pipeline, our next question will be, *which platform? Cloud or on-premise?*

To answer these questions, let's look at any constraints that we have. To analyze the streaming data, we must pull and read each customer's credit history record. However, since the credit history of a customer is sensitive information, we would prefer to use that information from an on-premise application. However, the company's mobile or web backend systems are deployed on the cloud. So, it makes sense to store the analyzed data on the cloud since it will take less time for the mobile or other web applications to fetch the data from the cloud than from on-premise. So, in this case, we will go with a hybrid approach, in which credit history data will be stored on-premise and the data will be analyzed and processed on-premise, but the resultant data will be stored in the cloud so that it can easily be retrieved from mobile and web backend systems.

In this section, we analyzed the data engineering problem and realized that this is a real-time stream processing problem, where the processing will happen on-premise. In the next section, we will use the result of this analysis and connect the dots to design the data pipeline and choose the correct technology stack.

Architecting the solution

To architect the solution, let's summarize the analysis we discussed in the previous section. Here are the conclusions we can make:

- This is a real-time data engineering problem
- This problem can be solved using a streaming platform such as Kafka or Kinesis
- 1 million events will be published daily, with a chance of the volume of events increasing over time
- The solution should be hosted on a hybrid platform, where data processing and analysis are done on-premise and the results are stored in the cloud for easy retrieval

Since our streaming platform is on-premise and can be maintained on on-premise servers, Apache Kafka is a great choice. It supports a distributed, fault-tolerant, robust, and reliable architecture. It can be easily scaled by increasing the number of partitions and provides an at-least-once delivery guarantee (which ensures that at least one copy of all events will be delivered without event drops).

Now, let's see how we will determine how the results and other information will be stored. In this use case, the credit history of an individual has a structured format and should be stored on-premise. RDBMS is a great option for such data storage. Here, we will be using PostgreSQL for this because PostgreSQL is open source, enterprise-ready, robust, reliable, and high performing (and also because we used it as an RDBMS option in *Chapter 4, ETL Data Load – A Batch-Based Solution to Ingesting*

Data in a Data Warehouse). Unlike credit history, the applications need to be accessed by mobile and web backends running on AWS, so the data storage should be on the cloud.

Also, let's consider that this data will primarily be consumed by mobile and web backend applications. So, would it be worth storing the data in a document format that can be readily pulled and used by the web and mobile backends? MongoDB Atlas on AWS cloud is a great option for storing documents in a scalable way and has a pay-as-you-go model. We will use MongoDB Atlas on AWS as the sink of the resultant data.

Now, let's discuss how we will process the data in real time. The data will be sent as events to a Kafka topic. We will write a streaming application to process and write the result event on an output topic. The resulting record will contain the risk score as well. To dump the data from Kafka to any other data store or database, we can either write a consumer application or use Kafka Sink connectors. Writing a Kafka consumer app requires development and maintenance effort. However, if we choose to use Kafka Connect, we have to just configure it to get the benefits of a Kafka consumer. Kafka Connect is faster to deliver, easier to maintain, and more robust as all exception handling and edge cases are already taken care of and well-documented. So, we will use a Kafka Sink connector to save the result events from the output topic to the MongoDB Atlas database.

The following diagram describes the solution architecture:

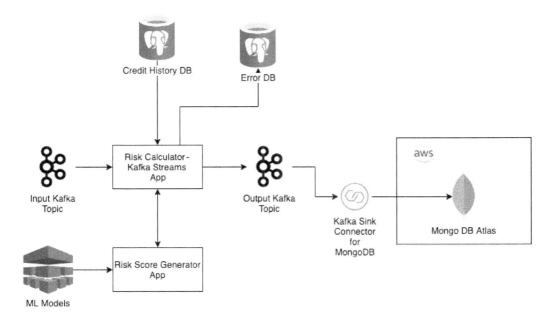

Figure 6.1 – Solution architecture for our real-time credit risk analyzer

As shown in the preceding diagram, our solution architecture is as follows:

- A new application event gets published in the input Kafka topic
- The Kafka Streams application – the Risk Calculator app – reads the application event and fetches the corresponding credit history of the applicant from the credit history database
- The Risk Calculator app creates and sends an HTTP request to the Risk Score Generator app with all the required parameters
- The Risk Score Generator app uses the already trained ML models to calculate the risk score of the application and returns the result to the Risk Calculator app
- The Risk Calculator app generates the enriched application event and writes the resultant event in the output topic
- A Kafka Sink connector, which is configured on the output topic, is responsible for consuming and writing the data to the MongoDB Atlas cloud database
- If there is a processing error during Kafka streaming, an error message, along with the input event, will be written in the error database

Now that we have learned how to architect a solution for our real-time data analysis needs, let's learn how to implement the architecture.

Implementing and verifying the design

The first step in a real-time implementation like this is to set up the streaming platform. To implement our architecture, we need to install Apache Kafka and create the necessary topics on our local machine.

Setting up Apache Kafka on your local machine

In this section, you will learn how to set up an Apache Kafka cluster, run it, and create and list topics. Follow these steps:

1. Download Apache Kafka version 2.8.1 from `https://archive.apache.org/dist/kafka/2.8.1/kafka_2.12-2.8.1.tgz`.

2. Extract the `kafka_2.12-2.8.1.tgz` archive file. The following command will help you do the same on Linux or macOS:

    ```
    $ tar -xzf kafka_2.12-2.8.1.tgz
    $ cd kafka_2.12-2.8.1
    ```

3. Navigate to the Kafka installation root directory and start zookeeper using the following command:

    ```
    $ bin/zookeeper-server-start.sh config/zookeeper.
    properties
    ```

4. Next, run the Kafka server using the following command:

    ```
    $ bin/kafka-server-start.sh config/server.properties
    ```

5. Next, create the topics using the following commands:

    ```
    $ bin/kafka-topics.sh --create --topic landingTopic1
    --bootstrap-server localhost:9092
    $ bin/kafka-topics.sh --create --topic enrichedTopic1
    --bootstrap-server localhost:9092
    ```

    For the sake of simplicity, we have defined one partition and set the replication factor as 1. But in a real production environment, the replication factor should be three or more. The number of partitions is based on the volume and velocity of data that needs to be processed and the optimum speed at which they should be processed.

6. We can list the topics that we created in the cluster using the following command:

    ```
    $ bin/kafka-topics.sh --describe --bootstrap-server
    localhost:9092
    ```

Now that we have installed Apache Kafka and created the topics that we need for the solution, you can focus on creating the credit records table and error table in a PostgreSQL instance installed on your local machine. The DDL and DML statements for these tables are available at https://github. com/PacktPublishing/Scalable-Data-Architecture-with-Java/tree/main/ Chapter06/SQL.

Reference notes

If you are new to Kafka, I recommend learning the basics by reading the official Kafka documentation: https://kafka.apache.org/documentation/#gettingStarted. Alternatively, you can refer to the book *Kafka, The Definitive Guide*, by *Neha Narkhede*, *Gwen Sharipa*, and *Todd Palino*.

In this section, we set up the Kafka streaming platform and the credit record database. In the next section, we will learn how to implement the Kafka streaming application to process the application event that reaches *landingTopic1* in real time.

Developing the Kafka streaming application

Before we implement the solution, let's explore and understand a few basic concepts about Kafka Streams. Kafka Streams provides a client library for processing and analyzing data on the fly and sending the processed result into a sink (preferably an output topic).

A stream is an abstraction that represents unbound, continuously updating data in Kafka Streams. A stream processing application is a program written using the Kafka Streams library to process data that is present in the stream. It defines processing logic using a topology. A Kafka Streams topology is a graph that consists of stream processors as nodes and streams as edges. The following diagram shows an example topology for Kafka Streams:

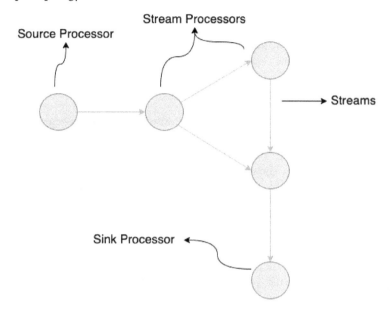

Figure 6.2 – Sample Kafka Streams topology

As you can see, a topology consists of **Stream Processors** – these are nodes and edges that represent streams. There can be two kinds of special stream processor nodes, as follows:

- **Source Processor**: This is a special stream processing node that produces an input stream of data from consuming messages from one or multiple Kafka topics
- **Sink Processor**: As the name suggests, a sink processor consumes data from upstream and writes it to a sink or target topic

A topology in a Kafka streaming application can be built using a low-level Processor API or using high-level **Domain-Specific Language** (DSL) APIs. When an event is published to a source Kafka topic, the topology gets triggered, which processes the event using the topology definition and publishes the processed event to the Sink topic. Once a topology is successfully invoked and completed on a source event, the event offset is committed.

In our use case, the Kafka Streams application will do the following:

- For the application event received, find the credit history from the credit record database.
- Create the ML request body using the data received from Kafka and the data pulled out from the credit record database

- Make a REST call to the Risk Score Generator application
- Form the final output record
- Send the final output record to a sink topic using a Sink processor

First and foremost, we need to create a Spring Boot Maven project and add the required Maven dependencies. The following Spring Maven dependencies should be added to the pom.xml file, as follows:

```
<dependency>
    <groupId>org.springframework.boot</groupId>
    <artifactId>spring-boot-starter-test</artifactId>
    <scope>test</scope>
</dependency>


<!-- https://mvnrepository.com/artifact/org.springframework.
boot/spring-boot-starter-jdbc -->
<dependency>
    <groupId>org.springframework.boot</groupId>
    <artifactId>spring-boot-starter-jdbc</artifactId>
</dependency>


<!-- https://mvnrepository.com/artifact/org.springframework.
boot/spring-boot-starter-web -->
<dependency>
    <groupId>org.springframework.boot</groupId>
    <artifactId>spring-boot-starter-web</artifactId>
</dependency>
```

Apart from this, as we are planning to develop a Kafka streaming application, we also need to add Kafka-related Maven dependencies, as follows:

```
<!-- Kafka dependencies -->


<dependency>
    <groupId>org.springframework.kafka</groupId>
    <artifactId>spring-kafka</artifactId>
    <version>2.6.2</version>
</dependency>
```

```xml
<dependency>
    <groupId>org.springframework.kafka</groupId>
    <artifactId>spring-kafka-test</artifactId>
    <version>2.6.2</version>
    <scope>test</scope>
</dependency>

<dependency>
    <groupId>org.apache.kafka</groupId>
    <artifactId>kafka-streams</artifactId>
    <version>3.0.0</version>
</dependency>
```

First, let's write the main class, where we will initialize the Kafka Spring Boot application. However, in our application, we must exclude KafkaAutoConfiguration (as we intend to use our own property names for Kafka-related fields and not Spring Boot's default Kafka property names), as shown in the following code:

```java
@SpringBootApplication(exclude = KafkaAutoConfiguration.class)
@Configuration
public class CreditRiskCalculatorApp {

    public static void main(String[] args) {
        SpringApplication.run(CreditRiskCalculatorApp.class);
    }
    . . .
}
```

After creating the main class, we will create the main KafkaStreamConfiguration class, where all streaming beans will be defined and instantiated. This is where we will use Kafka Streams DSL to build the topology. This class must be annotated with @EnableKafka and @EnableKafkaStreams, as shown in the following code snippet:

```java
@Configuration
@EnableKafka
@EnableKafkaStreams
public class KStreamConfiguration {
    . . .
```

Next, we will create the `KafkaStreamsConfiguration` bean. The following code snippet shows the implementation of the `KafkaStreamsConfiguration` bean:

```
@Bean(name = KafkaStreamsDefaultConfiguration.DEFAULT_STREAMS_
CONFIG_BEAN_NAME)
public KafkaStreamsConfiguration kStreamsConfig(){
    Map<String,Object> props = new HashMap<>();
    props.put(StreamsConfig.APPLICATION_ID_CONFIG,appId);
    props.put(StreamsConfig.BOOTSTRAP_SERVERS_
CONFIG,bootstrapServer);
    props.put(StreamsConfig.DEFAULT_KEY_SERDE_CLASS_CONFIG,
Serdes.String().getClass());
    props.put(StreamsConfig.DEFAULT_VALUE_SERDE_CLASS_
CONFIG,Serdes.String().getClass());
    props.put(StreamsConfig.DEFAULT_DESERIALIZATION_EXCEPTION_
HANDLER_CLASS_CONFIG, LogAndContinueExceptionHandler.class);

    return new KafkaStreamsConfiguration(props);
}
```

While creating the `KafkaStreamsConfiguration` bean, we must pass all Kafka streaming-related properties. Here, it is mandatory to set `StreamsConfig.APPLICATION_ID` and `StreamsConfig.BOOTSTRAP_SERVERS_CONFIG`. In this case, `StreamsConfig.APPLICATION_ID` corresponds to the consumer group ID of the Kafka Streams application, while `StreamsConfig.BOOTSTRAP_SERVERS_CONFIG` corresponds to the Kafka broker address. Without these values, no Kafka streaming or consumer application can run or connect to the Kafka cluster. Kafka Streams applications can distribute the traffic coming from a topic within a consumer group among multiple consumers that share the same consumer group ID. By increasing the running instance of the streaming application while using the same ID, we can have more parallelism and better throughput. However, increasing the number of instances beyond the number of partitions in the Kafka topic will not have any effect on the throughput.

Now that we have created the `KafkaStreamsConfiguration` bean, let's create `KStream`. While creating this `KStream` bean, we must define the topology. The following code creates the `KStream` bean:

```
@Bean
public KStream<String,String> kStream(StreamsBuilder builder){
    KStream<String,String> kStream = builder.
stream(inputTopic);
    kStream.transform(()->new RiskCalculateTransformer
```

```
(jdbcTemplate,restTemplate,mlRequestUrl)).to(outTopic);
    return kStream;
}
```

Each message in a Kafka topic consists of a key and a value. The value contains the actual message, while the key helps determine the partition while the message is published. However, when we consume the message using streams, we must mention the type of key and value that we are expecting. In our case, we are expecting both the key and value to be `String`. So, the `KStream` bean is created as an instance of `KStream<String, String>`. First, we must create a stream using the `StreamsBuilder` class, which is part of the Kafka Streams API. In our use case, the topology is built as follows:

1. First, using the `StreamsBuilder` API, input streams are created from `inputTopic`.
2. A transform processor is applied to the resultant input stream using the `transform()` DSL function.
3. A custom Transformer called `RiskCalculatorTransformer` is used to transform/process the data coming from the input stream.
4. The processed output event is written to `outputTopic`.

Now, let's learn how to write a custom Transformer for a Kafka Streams application. In our scenario, we have created `RiskCalculatorTransformer`. The following discussion explains how to develop a custom Transformer:

First, we must create a class that implements the `org.apache.kafka.streams.kstream.Transformer` interface. It has three methods – `init`, `transform`, and `close` – that need to be implemented. The following code shows the definition of the `Transformer` interface:

```
public interface Transformer<K, V, R> {
    void init(ProcessorContext var1);

    R transform(K var1, V var2);

    void close();
}
```

As you can see, the `Transformer` interface expects three generic types – K, V, and R. K specifies the data type of the key of the message, V specifies the data type of the value of the message, and R specifies the data type of the result of the message. While `init` and `close` are only used when some pre or post-processing is needed before the message is processed, `transform` is a mandatory method that defines the actual transformation or processing logic.

In our use case, we receive the value of the message as a JSON string, process it, add the risk score, and send out the resultant value as a JSON string. The data type of the key remains unchanged. Hence, we send out a KeyValue pair object as a result. Our final Transformer outline looks as follows:

```java
public class RiskCalculateTransformer implements
Transformer<String,String, KeyValue<String,String>> {

    @Override
    public void init(ProcessorContext processorContext) {
      ...
    }

    @Override
    public KeyValue<String, String> transform(String key,
String value) {
        ...
    }

    @Override
    public void close() {
      ...
    }
}
```

As shown in the preceding code, our Transformer is expecting the key and value of the message to be of the String type, and it returns a KeyValue pair where both the key and value are of the String type.

In our Transformer, we don't need any pre or post-processing. So, let's move on and discuss how to implement the transform method of our Transformer. The code of the transform method is as follows:

```java
@Override
public KeyValue<String, String> transform(String key, String
value) {
    try {
        ApplicationEvent event = mapper.
readValue(value,ApplicationEvent.class);
        List<CreditRecord> creditRecord = jdbcTemplate.
query(String.format("select months_balance,status from
chapter6.creditrecord where id='%s'",event.getId()),new
BeanPropertyRowMapper<CreditRecord>(CreditRecord.class));
```

```java
        MLRequest mlRequest = new MLRequest();
        mlRequest.setAmtIncomeTotal(event.getAmtIncomeTotal());
        ...


        HttpEntity<MLRequest> request = new
HttpEntity<>(mlRequest);
        ResponseEntity<RiskScoreResponse> response =
restTemplate.exchange(mlRequestUrl, HttpMethod.POST, request,
RiskScoreResponse.class);
        if(response.getStatusCode() == HttpStatus.OK){
            EnrichedApplication enrichedApplicationEvent = new
EnrichedApplication();
            enrichedApplicationEvent.
setApplicationforEnrichedApplication(event);
            enrichedApplicationEvent.setRiskScore(response.
getBody().getScore());
            return KeyValue.pair(key,mapper.
writeValueAsString(enrichedApplicationEvent));
        }else{
            throw new Exception("Unable to generate risk score.
Risk REST response - "+ response.getStatusCode());
        }

    } catch (Exception e) {
        ...
    }
    return null;
}
```

Here is the step-by-step guide for implementing our transform method:

1. First, we deserialize the incoming value, which is a JSON string, into a POJO called ApplicationEvent using the Jackson ObjectMapper class.

2. Then, we initiate a JDBC call to the credit record database using Spring's JdbcTemplate. While forming the SQL, we use the application ID that was deserialized in the previous step. We get a list of the CreditRecord objects because of the JDBC call.

3. Next, we form the request body for the HTTP REST call that we are going to make to get the risk score. Here, we populate an MLRequest object using the ApplicationEvent object (deserialized earlier) and the list of CreditRecord objects we obtained in the previous step.

4. Then, we wrap the `MLRequest` object in an `HTTPEntity` object and make the REST call using the Spring `RestTemplate` API.

5. We deserialize the REST response to the `RiskScoreResponse` object. The model of the `RiskScoreResponse` object looks as follows:

```
public class RiskScoreResponse {
    private int score;
    public int getScore() {
        return score;
    }
    public void setScore(int score) {
        this.score = score;
    }
}
```

6. If the REST response is `OK`, then we form the `EnrichedApplication` object using the `ApplicationEvent` and `RiskScoreResponse` objects.

7. Finally, we create and return a new `KeyValue` pair object, where the key is unchanged, but the value is the serialized string of the `EnrichedApplication` object we created in *step 6*.

8. For exception handling, we log any errors as well as send the error events to an error database for future analysis, reporting, and reconciliation. The reporting and reconciliation processes won't be covered here and are usually done by some kind of batch programming.

In this section, we learned how to develop a Kafka Streams application from scratch. However, we should be able to successfully unit test a streaming application to make sure that our intended functionalities are working fine. In the next section, we will learn how to unit test a Kafka Streams application.

Unit testing a Kafka Streams application

To unit test a Kafka Streams application, we must add the Kafka Streams test utility dependencies to the pom.xml file:

```
<!-- test dependencies -->
<dependency>
    <groupId>org.apache.kafka</groupId>
    <artifactId>kafka-streams-test-utils</artifactId>
    <version>3.0.0</version>
    <scope>test</scope>
</dependency>
<dependency>
```

```xml
        <groupId>junit</groupId>
        <artifactId>junit</artifactId>
        <version>4.12</version>
        <scope>test</scope>
    </dependency>
    <dependency>
        <groupId>org.hamcrest</groupId>
        <artifactId>hamcrest-core</artifactId>
        <version>1.3</version>
        <scope>test</scope>
    </dependency>
```

Also, before we do a JUnit test, we need to refactor our code a little bit. We have to break the definition of the `KStream` bean into two methods, like so:

```java
@Bean
public KStream<String,String> kStream(StreamsBuilder builder){
    KStream<String, String> kStream = StreamBuilder.INSTANCE.
getkStream(builder,inputTopic,outTopic,mlRequestUrl,
jdbcTemplate,restTemplate);
    return kStream;
}
...


public enum StreamBuilder {
    INSTANCE;
    public KStream<String, String> getkStream(StreamsBuilder
builder, String inputTopic,String outTopic, String
mlRequestUrl, JdbcTemplate jdbcTemplate, RestTemplate
restTemplate) {
        KStream<String,String> kStream = builder.
stream(inputTopic);
        kStream.transform(()->new RiskCalculateTransformer
(jdbcTemplate,restTemplate,mlRequestUrl)).to(outTopic);
        return kStream;
    }
}
```

As shown in the preceding code, we took out the KStream formation code, put it in a utility method in a singleton class called StreamBuilder, and used the Bean method as a wrapper on top of it.

Now, let's learn how to write the JUnit test case. First, our transformation requires a JDBC call and a REST call. To do so, we need to mock the JDBC call. To do that, we will use Mockito libraries.

We can mock our JdbcTemplate call like so:

```
@Mock
JdbcTemplate jdbcTemplate;
...
public void creditRiskStreams(){
    ...
    List<CreditRecord> creditRecords = new ArrayList<>();
    CreditRecord creditRecord = new CreditRecord();
    . . .
    creditRecords.add(creditRecord);
    Mockito. lenient().when(jdbcTemplate.
query("select months_balance,status from
chapter6.creditrecord where id='5008804'",new
BeanPropertyRowMapper<CreditRecord>(CreditRecord.class)))
            .thenReturn(creditRecords);
        ...
```

First, we create a mock JdbcTemplate object using the @Mock annotation. Then, we use Mockito's when().thenReturn() API to define a mock output for a call made using the mock JdbcTemplate object.

A similar technique can be used to mock RestTemplate. The code for mocking RestTemplate is as follows:

```
@Mock
private RestTemplate restTemplate;
public void creditRiskStreams(){
    ...


    RiskScoreResponse riskScoreResponse = new
RiskScoreResponse();
    ...
    Mockito
            .when(restTemplate.exchange(Mockito.anyString(),
```

```
HttpMethod.POST, Mockito.any(), RiskScoreResponse.class))
        .thenReturn(new ResponseEntity(riskScoreResponse,
HttpStatus.OK));
```

As you can see, first, we mock RestTemplate using the @Mock annotation. Then, using Mockito APIs, we mock any POST call that returns a RiskScoreResponse object.

Now, let's form the topology. You can use the following code to create the topology:

```
@Test
public void creditRiskStreamsTest() throws
JsonProcessingException {
    //test input and outputTopic
    String inputTopicName = "testInputTopic";
    String outputTopicName = "testOutputTopic";
    ...
    StreamsBuilder builder = new StreamsBuilder();
StreamBuilder.INSTANCE.
getkStream(builder,inputTopicName,outputTopicName,"any
url",jdbcTemplate,restTemplate);
    Topology testTopology = builder.build();
    ...
```

Here, we created an instance of the org.apache.kafka.streams.StreamsBuilder class. Using our StreamBuilder utility class, we defined the topology by calling the getkStream method. Finally, we built the topology by calling the build() method of the org.apache. kafka.streams.StreamsBuilder class.

Kafka Stream's test utils come with a Utility class called TopologyTestDriver. TopologyTestDriver is created by passing the topology and config details. Once TopologyTestDriver has been created, it helps to create TestInputTopic and TestOutputTopic. The following code describes how to instantiate a TopologyTestDriver and create TestInputTopic and TestOutputTopic:

```
public class CreditRiskCalculatorTests {
    private final Properties config;
public CreditRiskCalculatorTests() {
    config = new Properties();
    config.setProperty(StreamsConfig.APPLICATION_ID_CONFIG,
"testApp");
    config.setProperty(StreamsConfig.BOOTSTRAP_SERVERS_CONFIG,
"test:1234");
```

```
    config.setProperty(StreamsConfig.DEFAULT_KEY_SERDE_CLASS_
CONFIG, Serdes.String().getClass().getName());
    config.setProperty(StreamsConfig.DEFAULT_VALUE_SERDE_CLASS_
CONFIG, Serdes.String().getClass().getName());

}
. . .
@Test
public void creditRiskStreamsTest() throws
JsonProcessingException {
. . .
TopologyTestDriver testDriver = new
TopologyTestDriver(testTopology,config);
TestInputTopic<String,String> inputTopic = testDriver.
createInputTopic(inputTopicName, Serdes.String().serializer(),
Serdes.String().serializer());
TestOutputTopic<String,String> outputTopic = testDriver.
createOutputTopic(outputTopicName, Serdes.String().
deserializer(), Serdes.String().deserializer());

. . .
```

To create a `TestInputTopic`, we need to specify the name of the topic, as well as the key and value serializers. Similarly, `TestOutputTopic` requires key and value deserializers, along with the output topic name. We can push a test event to `TestInputTopic` using the following code:

```
inputTopic.pipeInput(inputPayload);
```

Finally, we can assert our expected result with the actual result using the `org.junit.Assert.assertEquals` static method, as follows:

```
assertEquals(mapper.readTree(outputTopic.readValue()),
mapper.readTree("{ \"id\": \"5008804\", \"genderCode\":
\"M\", \"flagOwnCar\": \"Y\", \"flagOwnRealty\": \"Y\",
\"cntChildren\": 0, \"amtIncomeTotal\": 427500.0,
\"nameIncomeType\": \"Working\", \"nameEducationType\":
\"Higher education\", \"nameFamilyStatus\": \"Civil
marriage\", \"nameHousingType\": \"Rented apartment\",
\"daysBirth\": -12005, \"daysEmployed\": -4542, \"flagMobil\":
1, \"flagWorkPhone\": 1, \"flagPhone\": 0, \"flagEmail\": 0,
\"occupationType\": \"\", \"cntFamMembers\": 2 , \"riskScore\":
3.0}"));
```

We can run this JUnit test by right-clicking and running the `Test` class, as shown in the following screenshot:

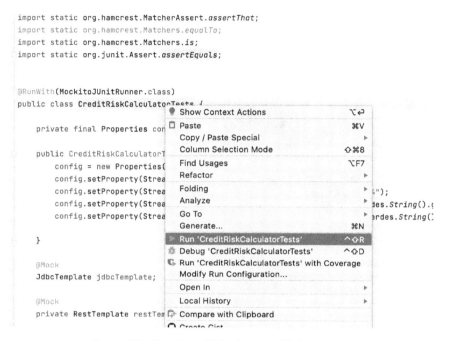

Figure 6.3 – Running a Kafka Streams JUnit test case

Once you have run the JUnit test case, you will see the test result in the run window of IntelliJ IDE, as shown in the following screenshot:

Figure 6.4 – Verifying the JUnit test's results

In this section, we learned how to write a JUnit test case for a Kafka streaming application and unit test our Streams application. In the next section, we will learn how to configure the streaming application and run the application on our local system.

Configuring and running the application

To run this application, we must configure the `application.yaml` file, which contains the following details:

- Application port number (as we will launch two Spring Boot applications on our local machine)
- Data source details
- Kafka details such as bootstrap servers and topics
- The REST HTTP URL for the Risk Score Generator app

Our sample `application.yaml` file will look as follows:

```
spring:
  datasource:
    url: jdbc:postgresql://localhost:5432/database
    username: postgres
    driverClassName: org.postgresql.Driver
riskcalc:
  bootstrap-servers: localhost:9092
  appId: groupId1
  inputTopic: landingTopic1
  outTopic: enrichedTopic1
  mlRequestUrl: "http://localhost:8081/riskgenerate/score"
```

Now, we can run the application by running the main class, `CreditRiskCalculatorApp`, of the CreditRiskCalculator application. But before we start the CreditRiskCalculator app, we should run the RiskScoreGenerator app by running its main class – that is, `RiskScoreGenerator`. Both these applications are Spring Boot applications; please refer to the *Implementing and unit testing the solution* section of *Chapter 4, ETL Data Load – A Batch-Based Solution to Ingesting Data in a Data Warehouse*, to learn how to run a Spring Boot application.

Troubleshooting tips

If, while starting the CreditRiskCalculator application, you notice a warning message such as **Connection to node -1 (localhost/127.0.0.1:9092) could not be established. Broker may not be available.** in the logs, please ensure your Kafka server is reachable and running.

If you notice an exception such as **org.apache.kafka.clients.consumer. CommitFailedException:Commit cannot be completed since the group has already rebalanced and assigned the partitions to another member**, then try to increase `max.poll.interval. ms` or decrease the value of `max.poll.records`. This usually happens when the number of records polled takes more time to process than the maximum poll interval time configured.

> If you encounter an error such as **java.lang.IllegalArgumentException: Assigned partition x-topic for non-subscribed topic regex pattern; subscription pattern is y-topic** while starting your streaming app, then check for other streaming or consumer applications that are using the same `application.id`. Change your `application.id` to solve this problem.

In this section, we learned how to create and unit test a Kafka Streams application. The source code for this application is available on GitHub at `https://github.com/PacktPublishing/Scalable-Data-Architecture-with-Java/tree/main/Chapter06/sourcecode/CreditRiskCalculator`.

Since implementing an ML-based Risk Score Generator app is outside the scope of this book, we have created a Spring Boot REST application that generates a dummy risk score between 1 to 100. The code base of this dummy application is available on GitHub at `https://github.com/PacktPublishing/Scalable-Data-Architecture-with-Java/tree/main/Chapter06/sourcecode/RiskScoreGenerator`.

In a real-world scenario, ML-based applications are more likely to be written in Python than Java since Python has better support for AI/ML libraries. However, a Kafka Streams application will be able to make a REST call and get the generated risk score from that application, as shown earlier.

So far, we have been receiving the event from an input Kafka topic, processing it on the fly, generating a risk score, and writing the enriched event to an output Kafka topic. In the next section, we will learn how to integrate Kafka with MongoDB and stream the events to MongoDB as soon as they are published to the output Kafka topic.

Creating a MongoDB Atlas cloud instance and database

In this section, we will create a cloud-based instance using MongoDB Atlas. Follow these steps to set up the MongoDB cloud instance:

1. Sign up for a MongoDB Atlas account if you haven't done so already (`https://www.mongodb.com/atlas/database`). While signing up, you will be asked for the type of subscription that you need. For this exercise, you can choose the shared subscription, which is free, and choose an AWS cluster as your preferred choice of cloud.

2. You will see the following screen. Here, click the **Build a Database** button to create a new database instance:

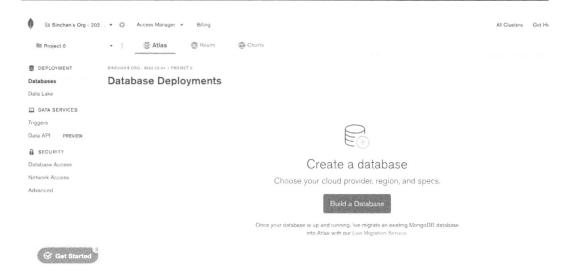

Figure 6.5 – MongoDB Atlas welcome screen

3. To provision a new database, we will be asked to set a username and a password, as shown in the following screenshot:

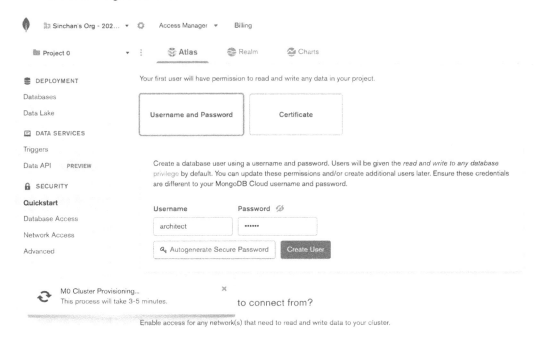

Figure 6.6 – Provisioning a new database instance

4.   Then, we will be asked to enter all the IP addresses that we want to grant access to the MongoDB instance. Here, since we will run our application from our local system, we will add our local IP address to the IP Access List. Then, click **Finish and Close**:

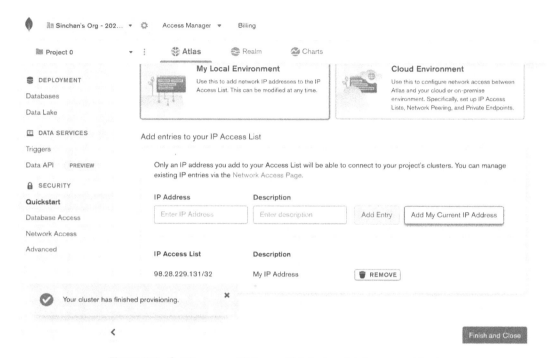

Figure 6.7 – Setting up an IP Access List during database provisioning

5.   Once the cluster has been created, we will see the cluster on the dashboard, as shown in the following screenshot. Now, click the **Browse Collections** button to see the collections and data in this database instance:

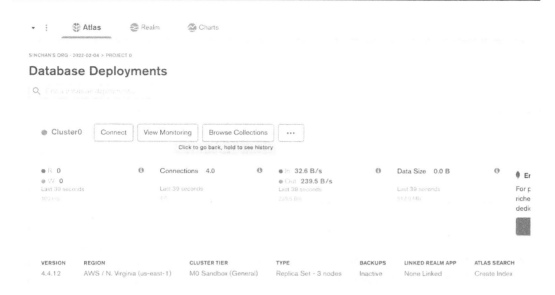

Figure 6.8 – Cluster dashboard in MongoDB Atlas

6. As shown in the following screenshot, currently, there are no collections or data. However, you can create collections or data manually while using this interface by clicking the **Add My Own Data** button:

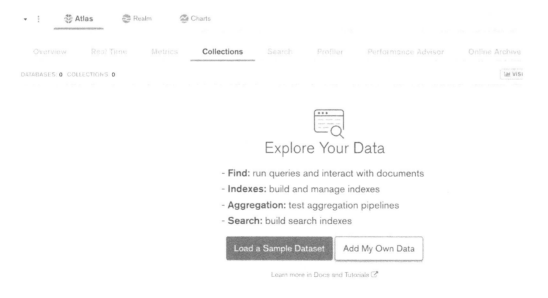

Figure 6.9 – Exploring collections and data in the MongoDB database instance

In this section, we learned how to create a cloud-based MongoDB instance using the online interface of MongoDB Atlas. In the next section, we will learn how to configure and deploy our MongoDB Kafka connectors to send the data from Kafka to MongoDB in real time.

Configuring Kafka Connect to store the results in MongoDB

Kafka Connect is an open source, pluggable data integration framework for Kafka. It enables data sources and data sinks to easily connect with Kafka. Instead of writing cumbersome code to publish a message from the data source or consume a message from Kafka to write into a data sink, Kafka Connect provides declarative configuration to connect to a data source or sink.

A Kafka Connect cluster already ships with a few types of connectors, such as `FileSourceConnector`. However, we can install any available connectors by placing them in the `plugins` folder. For our use case, we will deploy the MongoDB connector plugin (discussed later in this chapter).

A Kafka Connect instance can be deployed and run in either cluster or standalone mode. However, in production, it usually runs in cluster mode. When we run in cluster mode, we can register the Kafka connector configuration using Kafka Connects' REST API. In standalone mode, we can register a connector configuration while starting the Kafka Connect instance.

Since we are running our Kafka cluster on our local machine, we will deploy our Kafka Connect instance in standalone mode for this implementation. But remember, if you are implementing for production purposes, you should run Kafka, as well as Kafka Connect, in a clustered environment (this can be a physical cluster or a virtual cluster, such as a virtual machine, AWS ECS, or Docker container).

First, let's set up the Kafka Connect cluster:

1. First, create a new folder called `plugins` under the Kafka root installation folder, as shown in the following screenshot:

Figure 6.10 – Creating the plugins folder

2. Next, navigate to `connect-standalone.properties`, which is present in the `<Kafka-root>/config` folder. Add the following property to the `connect-standalone.properties` file:

    ... .

```
plugin.path=/<full path of Kafka installation root>/
plugins
```

3. Then, download the MongoDB Kafka connector plugin from `https://www.confluent.io/hub/mongodb/kafka-connect-mongodb`.

    A ZIP file will be downloaded. Copy and extract the ZIP file in the `plugin` folder created under the Kafka root installation folder. At this point, the folder structure should look as follows:

Figure 6.11 – Kafka folder structure after deploying the mongo-kafka-connect plugin

Now, let's learn how to create and deploy a Kafka Connect worker configuration to create a pipeline between a Kafka topic and MongoDB sink.

To write a Kafka Connect worker, we must understand the various types of declarative properties that Kafka Connect supports. The following diagram depicts various kinds of components that a Kafka Connect worker consists of:

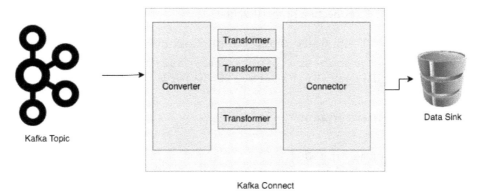

Figure 6.12 – Kafka Connect components

A Kafka connector consists of three types of components. They are as follows:

- **Connector**: This interfaces Kafka with external data sources. It takes care of implementing whatever external protocol those data sources and sinks need to communicate with Kafka.

- **Converter**: Converters are used to serialize and deserialize events.

- **Transformer**: This is an optional property. It is a stateless function that's used to slightly transform the data so that it is in the right format for the destination.

For our use case, we don't need a transformer, but we do need to set all the properties related to the connector and converter. The following code is for the Kafka Sink Connect worker:

```
name=mongo-sink
topics=enrichedTopic1
connector.class=com.mongodb.kafka.connect.MongoSinkConnector
tasks.max=1

# converter configs
key.converter=org.apache.kafka.connect.storage.StringConverter
value.converter=org.apache.kafka.connect.json.JsonConverter
key.converter.schemas.enable=false
value.converter.schemas.enable=false
...
```

As shown in the preceding configuration code, the connector properties such as `connector.class` and other properties specific to MongoDB are configured, and the converter properties such as `key.converter` and `value.converter` are set. Next, in the sink connector configuration, we define all the MongoDB connection properties, like so:

```
# Specific global MongoDB Sink Connector configuration
connection.uri=mongodb+srv://username:password@cluster0.ipguv.
mongodb.net/CRRD?retryWrites=true&w=majority
database=CRRD
collection=newloanrequest
max.num.retries=1
retries.defer.timeout=5000
```

Now, we will set the `document.id` and `writemodel.strategy` properties in the sink connector configuration, as shown here:

```
document.id.strategy=com.mongodb.kafka.connect.sink.processor.
id.strategy.PartialValueStrategy
document.id.strategy.partial.value.projection.list=id
```

```
document.id.strategy.partial.value.projection.type=AllowList
writemodel.strategy=com.mongodb.kafka.connect.sink.writemodel.
strategy.ReplaceOneBusinessKeyStrategy
```

Save this configurations in a property file called `connect-riskcalc-mongodb-sink. properties` and place it in Kafka Connect's `config` folder.

Now, we can run the Kafka Connect instance in standalone mode and start the `mongodb-sink` connector using the following command:

```
bin/connect-standalone.sh config/connect-standalone.properties
connect-riskcalc-mongodb-sink.properties
```

Now, let's learn how to troubleshoot possible issues that we may encounter.

Troubleshooting the Kafka Sink connector

When a source or a sink connector runs on a Kafka Connect cluster, you may encounter multiple issues. The following list specifies a few common issues and how to resolve them:

- If you encounter an error similar to the following, then please check whether the JSON message should contain a schema or not. If the JSON message should not contain schema, make sure that you set the `key.converter.schemas.enable` and `value.converter.schemas. enable` properties to `false`:

  ```
  org.apache.kafka.connect.errors.DataException:
  JsonConverter with schemas.enable requires "schema" and
  "payload" fields and may not contain additional fields.
  ```

- If you encounter an error such as `org.apache.kafka.common.errors. SerializationException: Error deserializing Avro message for id -1`, then check if the payload is Avro or JSON. If the message is a JSON payload instead of Avro, please change the value of the `value.converter` property in the connector to `org.apache.kafka.connect.json.JsonConverter`.

- You may encounter `BulkWriteExceptions` while writing to MongoDB. `BulkWriteExceptions` can be a `WriteError`, a `WriteConcernError`, or a `WriteSkippedError` (due to an earlier record failing in the ordered bulk write). Although we cannot prevent such errors, we can set the following parameters to move the rejected message to an error topic called `dead-letter-queue`:

  ```
  errors.tolerance=all
  errors.deadletterqueue.topic.name=<name of topic to use
  as dead letter queue>
  errors.deadletterqueue.context.headers.enable=true
  ```

In this section, we successfully created, deployed, and ran the MongoDB Kafka Sink connector. In the next section, we will discuss how to test the end-to-end solution.

Verifying the solution

To test the end-to-end pipeline, we must make sure that all the services, such as Kafka, PostgreSQL, Kafka Connect, and the MongoDB instance, are up and running.

Apart from that, the Kafka Streams application and the Risk Score Generator REST application should be up and running. We can start these applications by running the main Spring Boot application class.

To test the application, open a new Terminal and navigate to the Kafka root installation folder. To start an instance of the Kafka console producer, use the following command:

```
bin/kafka-console-producer.sh --topic landingTopic1
--bootstrap-server localhost:9092
```

Then, we can publish input messages by using the console producer, as shown in the following screenshot:

Figure 6.13 – Publishing messages using the Kafka console producer

As soon as we publish the message in the input topic, it gets processed, and the result is written to the MongoDB instance. You can verify the results in MongoDB like so:

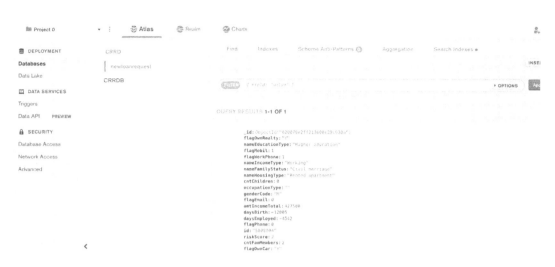

Figure 6.14 – Verifying the results

In this section, we learned how to test the end-to-end solution for a real-time data processing problem and verify the result. Now, let's summarize what we learned in this chapter.

Summary

In this chapter, we discussed how to analyze a real-time data engineering problem, identify the streaming platform, and considered the basic characteristics that our solution must have to become an effective real-time solution. First, we learned how to choose a hybrid platform to suit legal needs as well as performance and cost-effectiveness.

Then, we learned how to use our conclusions from our problem analysis to build a robust, reliable, and effective real-time data engineering solution. After that, we learned how to install and run Apache Kafka on our local machine and create topics in that Kafka cluster. We also learned how to develop a Kafka Streams application to do stream processing and write the result to an output topic. Then, we learned how to unit test a Kafka Streams application to make the code more robust and defect-free. After that, we learned how to set up a MongoDB Atlas instance on the AWS cloud. Finally, we learned about Kafka Connect and how to configure and use a Kafka MongoDB Sink connector to send the processed event from the output topic to the MongoDB cluster. While doing so, we learned how to test and verify the real-time data engineering solution that we developed.

With that, we have learned how to develop optimized and cost-effective solutions for both batch-based and real-time data engineering problems. In the next chapter, we will learn about the various architectural patterns that are commonly used in data ingestion or analytics problems.

7

Core Architectural Design Patterns

In the previous chapters, we learned how to architect data engineering solutions for both batch-based and real-time processing using specific use cases. However, we haven't discussed the various options available concerning architectural design patterns for batch and real-time stream processing engines.

In this chapter, we will learn about a few commonly used architectural patterns for data engineering problems. We will start by learning about a few common patterns in batch-based data processing and common scenarios where they are used. Then, we will learn about various streaming-based processing patterns in modern data architectures and how they can help solve business problems. We will also discuss the two famous hybrid data architectural patterns. Finally, we will learn about various serverless data ingestion patterns commonly used in the cloud.

In this chapter, we will cover the following topics:

- Core batch processing patterns
- Core stream processing patterns
- Hybrid data processing patterns
- Serverless patterns for data ingestion

Core batch processing patterns

In this section, we will look at a few commonly used data engineering patterns to solve batch processing problems. Although there can be many variations of the implementation, these patterns are generic, irrespective of the technologies used to implement the patterns. In the following sections, we'll discuss the commonly used batch processing patterns.

The staged Collect-Process-Store pattern

The **staged Collect-Process-Store pattern** is the most common batch processing pattern. It is also commonly known as the **Extract-Transform-Load** (ETL) pattern in data engineering. This architectural pattern is used to ingest data and store it as information. The following diagram depicts this architectural pattern:

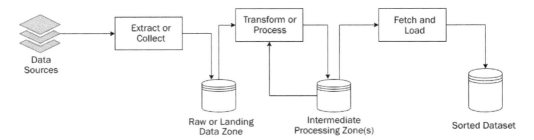

Figure 7.1 – The staged Collect-Process-Store pattern

We can break this pattern into a series of stages, as follows:

1. In this architectural pattern, one or more data sources are extracted and kept in a form of data storage called a raw zone or landing zone. The landing zone data is often raw data, which consists of noise such as extra spaces, junk characters, important fields missing, and so on. The extraction or collection job has the responsibility to extract and store the data in the raw zone. The data storage that's used for the landing zone can vary from a filesystem, the **Hadoop Distributed File System** (**HDFS**), an S3 bucket, or some relational database based on the use case and the platform chosen to solve the problem.

2. The processing jobs read the data from the raw zone and perform a series of transformations on the data, such as data cleansing, data standardization, and data validation. The job stores its output in the intermediate processing zone(s). There can be one or more transformation jobs, as well as multiple intermediate processing zones, based on the project and technology. Sometimes, processing jobs fetch related information from the intermediate data zone to enrich the processed data. In such a scenario, it reads the data from the intermediate processing zone or any external reference database. The final intermediate processing zone contains the data, which is cleansed, transformed, validated, and well organized.

3. The fetch and load process picks up the transformed data and loads it into the sorted dataset layer. The sorted dataset layer contains clean and usable data in a specific format that can easily be consumed by downstream applications for data analytics, reference, and so on. The sorted dataset layer is also popularly known as the **Organized Data Layer** (**ODL**). There is no hard and fast rule regarding the type of database or data store used for the sorted dataset layer. However, based on whether the sorted data will be used for **Online Transaction Processing** (**OLTP**) or **Online Analytical Processing** (**OLAP**), a database is chosen. Generally, this pattern is used to ingest and store data for OLAP purposes.

The jobs for this architectural pattern typically run periodically based on a predetermined schedule, such as once daily or once weekly, or every Friday and Wednesday at 8 P.M. One of the advantages of this pattern is that it ingests the data and processes in a series of stages. The output of each stage is stored in an intermediate processing zone, and the next stage fetches data from the output of the previous stage. This staged architecture makes the design loosely coupled. Often, in production, the data processing job fails. In such a situation, we don't need to rerun the full ingestion pipeline; instead, we can restart the pipeline from the job that failed.

Now, let's look at a real-world use case where this pattern will be a good fit. A health insurance firm receives tons of insurance claims every day. To process these claims and determine the cost that will be paid by the insurance firm, the data needs to be cleansed, enriched, organized, and stored. In such a use case, this architectural pattern can be used to ingest different kinds of claims, such as medical, dental, and vision, from various sources; then, they can be extracted, transformed, and loaded into ODL. Another example implementation of this pattern was discussed in *Chapter 4, ETL Data Load – A Batch-Based Solution to Ingesting Data in a Data Warehouse*.

Common file format processing pattern

Suppose there is a scenario where there are multiple files (say, for example, 25 source files) for the data sources and the structure of these sources are quite different from each other. Now, the question is, *Can the staged collect-process-store pattern handle such a use case?* Yes, it can. But is it optimized to do so? No, it's not. The problem is that for all 25 different kinds of source files, we need to have a separate set of transformation logic written to process and store them into a sorted dataset. We may require 25 separate data pipelines to ingest the data. This not only increases development effort but also increases analysis and testing effort. Also, we may need to fine-tune all the jobs in all 25 data pipelines. The **common file format processing pattern** is well suited to overcome such problems. The following diagram depicts how the common file format processing pattern works:

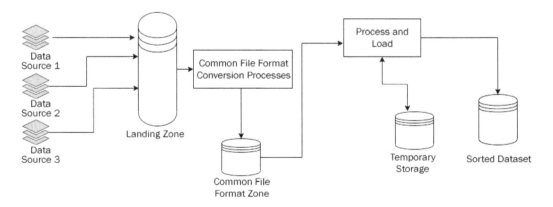

Figure 7.2 – The common file format processing pattern

This pattern is divided into the following stages:

1. In this architectural pattern, multiple source files with distinctively different source file structures are stored or sent to the landing zone from the sources. The landing zone can be a filesystem, a NAS mount, an SFTP location, or an HDFS location.

2. A common file format conversion process runs, which takes the different incoming source files and converts them into a uniform structure called a **common file format**. The job or pipeline that runs to do this conversion is lightweight. It should not do cleansing or business transformation in this layer. The common file format conversion process writes its output in the common file format zone.

3. Now that all the files are in the same format or structure, a single set of process and load jobs can run on top of those files present in the common file format zone. The process and load process can be a single job or a series of jobs that writes the final organized and sorted data into ODL or the sorted dataset layer. The process and load job may write its intermediate result to temporary storage zones if required.

Now, let's look at a real-world scenario. A credit card company wants to generate and provide offers for its customers based on their buying and spending patterns, as well as a set of complex rules. However, transaction data can be received from various kinds of sources, which includes web-based payment gateways, physical transaction gateways, payment apps such as PayPal and Cash App, foreign payment gateways, and various other similar apps. However, the transaction files that are received from all these sources are in different formats. One option is to create a separate set of transformation mappings for each source and apply the rules differently. However, that will cause a lot of development time and costs, as well as a maintenance challenge. In such a scenario, the common file format processing pattern can be used to convert all transaction files coming from different source systems into a common file format. Then, a single set of rule engine jobs can process transactions received from different sources.

The Extract-Load-Transform pattern

Previously in this book, we learned about the classical ETL-based pattern, where we extract the data first, transform and process the data, and finally store it in the final data store. However, with modern processing capabilities and the scalability that the cloud offers, we have seen many **Massive Parallel Processing** (MPP) databases such as Snowflake, Redshift, and Google's Big Query becoming popular. These MPP databases have enabled a new pattern of data ingestion where we extract and load the data into these MPP databases first and then process the data. This pattern is commonly known as the **Extract-Load-Transform** (**ELT**) pattern or the Collect-Store-Process pattern. This pattern is useful for building high-performing data warehouses that contain a huge amount of data. The following diagram provides an overview of the ELT pattern:

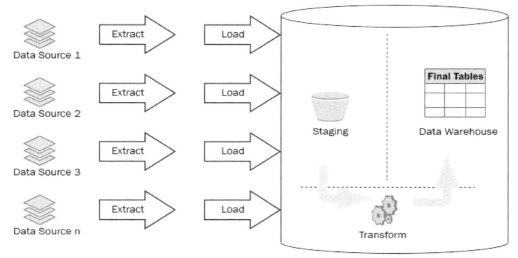

Figure 7.3 – The Extract-Load-Transform (ELT) pattern

The preceding diagram depicts the typical flow of an ELT pattern. This can be described as follows:

1. As shown in the preceding diagram, raw data is extracted and loaded into the MPP database. This data is stored in the staging zone of the MPP database.

2. Then, using MPP queries and a transformation pipeline, the data is transformed into the final set of tables. These final tables are exposed as a data warehouse. For security purposes, sometimes, views are created and exposed as a data warehouse on top of the tables.

Again, let's look at an example of how this pattern is used in the industry. As the customer experience continues to rise, businesses face a gap between the data needed to meet customer expectations and the ability to deliver using the current data management practice. Customer 360 involves building a complete and accurate repository of all the structured and unstructured data across the organization related to the customer. It is an aggregation of all customer data into a single unified location so that it can be queried and used for analytics to improve customer experience. To build Customer 360 solutions, we can leverage the power of MPP databases to create a single unified Customer 360 data warehouse. An example of a Customer 360 design using Snowflake on AWS is shown here:

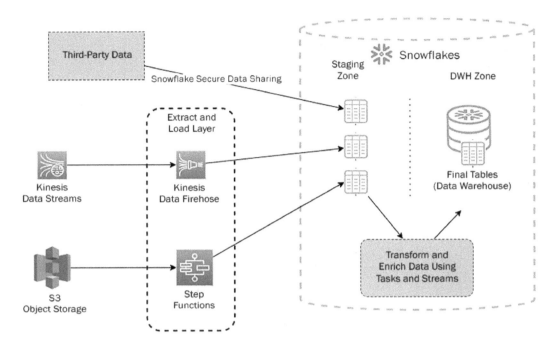

Figure 7.4 – Example of an ELT pattern on AWS

Here, data from cloud storage, event streams, and third-party sources all land in the staging area of Snowflake (an MPP database). Then, using Snowflake pipelines, data is cleansed, transformed, and enriched and is stored in the final tables to be consumed by the organization as the centralized enterprise data warehouse.

The compaction pattern

Data warehouses are not only built on MPP databases. For big data needs, a lot of the time, they are built on top of HDFS using Hive as the querying engine. However, in modern pipelines, a lot of data is dumped in the landing zone by real-time processing engines such as Kafka or Pulsar. Although the use case needs our processing jobs to run a few times a day or once daily, the files are landed when any records come in. This creates a different kind of problem. Due to the scenario described earlier, too many small files containing few records are created. HDFS is not designed to work with small files, especially if it is significantly smaller than the HDFS block size; for example, 128 MB. HDFS works better if a smaller number of huge files are stored instead of a huge number of small files.

Eventually, as the small files grow, the query performance reduces, and eventually, Hive is unable to query those records. To overcome this problem, a pattern is commonly used. This is called the compaction pattern. The following diagram provides an overview of the compaction pattern:

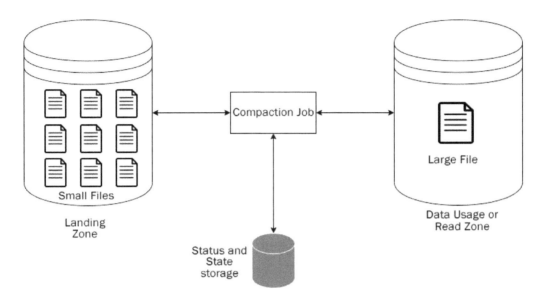

Figure 7.5 – The compaction pattern

In this architectural pattern, the small files are stored in the landing zone. A batch-based periodical job runs and compacts those small files to create a single large file. In between, it uses the status and state storage to store job audit information. It is also used to store state information that may be used by subsequent compaction jobs.

The staged report generation pattern

We have discussed multiple patterns to show how data is ingested and stored as a sorted dataset or in a data warehouse. This pattern, on the other hand, focuses on running data analytics jobs and generating report(s) from the **ODL** or data warehouse. The following diagram shows the generic architecture of this pattern:

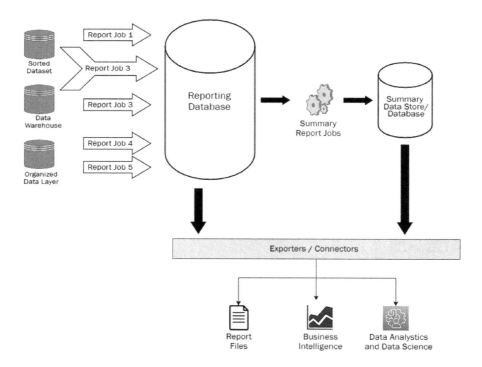

Figure 7.6 – The staged report generation pattern

The staged report generation pattern consists of primarily two stages and an auxiliary step, as follows:

1. **Report generation stage**: Various analytics jobs run on top of the sorted data or organized data layer. Such analytics jobs may even run on the data stored in the data warehouse. These jobs then save the report of the analysis in a reporting database. A reporting database can be a relational database, a NoSQL database, or a search engine such as Elasticsearch.

2. **Summary generation stage**: The summary reporting jobs fetch data from the reporting database and report the summary data in the summary database. Summary databases are usually relational databases, data warehouses, or search engines.

3. Using exporters and connectors, the data present in either the reporting database or the summary database can be visualized using BI tools or used for data science and analytics purposes, or simply used to extract flat files containing reports.

Now, let's look at a real-world scenario where this pattern is suitable. Let's say that a company has an on-premises data center. Every day, monitoring and resolution logs are generated for all the servers and storage, backup storage, and networking devices present in the data center. This data is ingested and stored in a data warehouse that contains daily, weekly, and monthly outage and resolution details. Using this data warehouse, the organization wants to generate various reports for the average SLA for various kinds of incidents, the performance or KPI ratio before and after the resolutions, and the team-wise velocity of closing incidents.

Finally, the company wants to generate a summary of all incidents on a weekly, monthly, and quarterly basis. This use case is well-suited for using this pattern. In this use case, we can generate all the reports and store them in a reporting database and generate the summary reports to the summary database. Both general reports and summary reports can be visualized using BI tools such as Tableau by pulling the data from the reporting databases using proper connectors.

In this section, we learned about a few popular batch processing architectural patterns and a few real-world scenarios that can be applied. In the next section, we will cover a few common patterns used for real-time stream processing.

Core stream processing patterns

In the previous section, we learned about a few commonly used batch processing patterns. In this section, we will discuss various stream processing patterns. Let's get started.

The outbox pattern

With modern data engineering, monolithic applications have been replaced by a series of microservices application working in tandem. Also, it is worth noting that microservices usually don't share their databases with other microservices. The database session commits and interservice communications should be atomic and in real time to avoid inconsistencies and bugs. Here, the outbox pattern comes in handy. The following diagram shows the generic architecture of the outbox pattern:

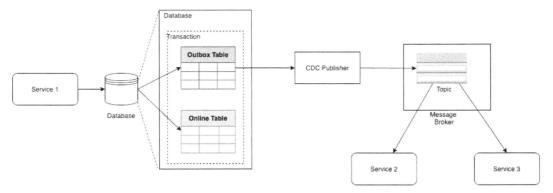

Figure 7.7 – The outbox pattern

As we can see, a microservice (here, **Service 1**) writes a transaction to not only the required table where online reads and writes happen (denoted in the diagram as **Online Table**) but also to an **Outbox Table**, whose structure is where messages to the message broker should be published. Just like the physical trays on office desks that once held outgoing letters and documents, the outbox pattern uses an **Outbox**

Table to send messages to the message broker. A **Change Data Capture (CDC)** publisher picks the CDC events from the **Outbox Table** area and publishes them to our **Message Broker**. Downstream services that need data from Service 1 consume the data from the topic.

The saga pattern

The saga pattern is a design pattern that is used to manage and handle distributed transactions across multiple applications or services successfully. In a real-world scenario, a single business transaction can never be done with one application or backend service. Usually, multiple applications work in tandem to complete a successful business transaction. However, we need to have an asynchronous, reliable, and scalable way to communicate between these systems. Each business transaction that spans multiple services is called a saga. The pattern to implement such a transaction is called the saga pattern.

To understand the saga pattern, let's take a look at an e-commerce application. The following diagram shows the workflow of a simplified ordering system in an e-commerce application:

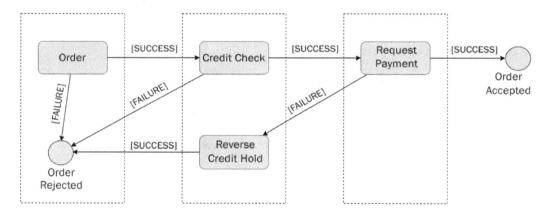

Figure 7.8 – A simplified e-commerce ordering system

As we can see, an ordering system consists of multiple services, each of which has its own set of functions to perform. Essentially, there are three services: the ordering service, the credit management service, and the payment service. For a successful ordering transaction, the ordering service receives the order. If the order is received successfully, it goes to the credit management service, which checks the credit card's balance and validates the card. If the credit check is successful, the system uses the payment service to request payment. If the payment goes through successfully, the order is marked as accepted. If it fails at any stage, the transaction is aborted, and the order gets rejected.

Now, let's see how the saga pattern is implemented in this situation. Here inter-service communication is decoupled and made asynchronous by introducing a message broker platform to exchange messages between them. The following diagram shows how the saga pattern is used to implement the ordering system for an e-commerce application:

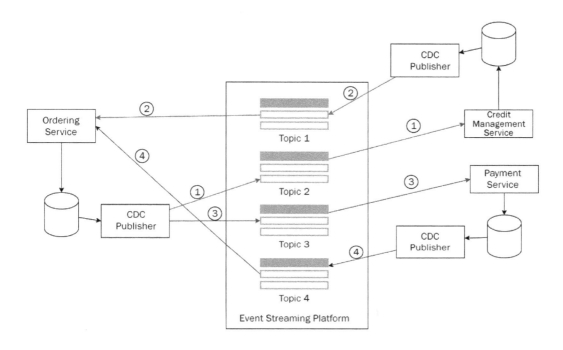

Figure 7.9 – The saga pattern applied to implement an ordering system

Here, the saga pattern is applied to the saga transaction of placing an order. Here, the ordering service stores the orders in a local database. A CDC publisher is used to publish the messages containing the order to **Topic 2** in the streaming platform. Data sent to Topic 2 is consumed by the credit management service to do the credit check functionality (marked as flow *1* in the preceding diagram). The output of the credit check functionality is sent to a local database. The message containing the credit check result is sent from the local database to **Topic 1**. The ordering service consumes and stores the output for further processing.

If the credit check report is positive, a payment request event is published in **Topic 3** using the CDC processor (depicted as flow *3* in the preceding diagram). The event that's published in flow *3* is picked up by the payment service and requests payment. The result of the payment request is saved in the local payment database. The CDC publisher from the payment database produces the payment output to **Topic 4**, which is denoted as flow *4* in the preceding diagram. Using the information that's been shared over **Topic 4**, the ordering service determines whether the order was placed or whether it was rejected. One of the interesting things that you can see is that each step of the saga pattern follows the outbox pattern, as described earlier. We can say a series of outbox patterns are knitted in a certain manner to create the saga pattern.

The choreography pattern

This pattern is used specifically where each component independently takes part in the decision-making process to complete a business transaction. All these independent components talk to a centralized orchestrator application or system. Just like how in choreography, the choreography pattern enables all independent dancers to perform separately and create a wonderfully synchronized show, so an orchestrator orchestrates decentralized decision-making components to complete a business transaction. This is the reason that this pattern is called the choreography pattern. The following diagram provides an overview of the choreography pattern:

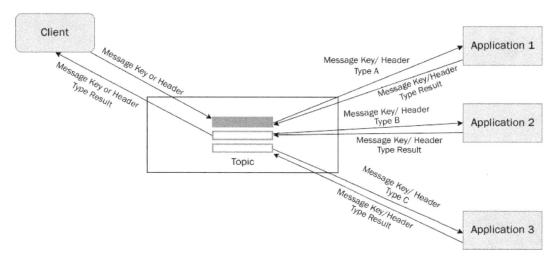

Figure 7.10 – The choreography pattern

As we can see, events from the client are streamed to the topic. Each event contains a specific message header or a message key. Based on the type of message header value or key value, each consuming application can filter and process the messages required by that application. Once it processes the event, it generates a result event to the same topic but with a different key or header value. The client consumes all the resulting events to create the final output or decision.

This pattern is useful when you have scenarios where applications may be frequently added, removed, or updated or there is a bottleneck in the centralized orchestration layer.

Let's take a look at a real-world use case where this pattern may come in handy. A service provider receives different events whenever a client does a recharge. In a single recharge, the client can buy different bundles, such as a top-up bundle, data bundle, and so on. Each bundle adds a message header to the event. Different client applications provide customized offers for each kind of bundle. This use case is suitable for the choreography pattern. Suppose an event comes with both the top-up and data bundles; this will add two pieces of header information, so there will be two consuming applications based on the type of bundle that will be consumed; its own set of offers will be generated and sent back to the client using the topic. It makes sense to use the choreography pattern here as the type of bundles are dynamic, which can vary year-to-year and season-to-season. So, the consuming applications may be frequently added or removed from the ecosystem.

The Command Query Responsibility Segregation (CQRS) pattern

This is a very famous pattern, where the read responsibility and write responsibility is segregated out. This means the data is written to a different data store and read from another data store. While the write data store is optimized for fast writes, the read data store is optimized for fast data reads. The following diagram shows how the CQRS pattern works:

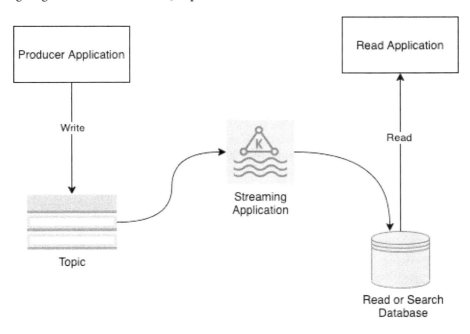

Figure 7.11 – The CQRS pattern

The preceding diagram depicts how the CQRS pattern works. The flow of this pattern is as follows:

1. First, the producer or publisher writes the event to a topic.

2. Using a streaming application, this record is streamed into the read database. While the topic is optimized for fast data writes, the read database is optimized for high-performance data reads. This kind of pattern is very useful for scenarios where we need to have a high write as well as high read speed.

For example, for a big e-commerce website such as Amazon, the traffic increases heavily on Amazon sale days. In this scenario, there is the possibility of a high number of writes as well as a high number of searches. In this case, various sources such as mobile apps, web portals, and so on will accept orders and update the inventory. Also, offers and discounts are changed hourly using the Amazon Big Day sale event management portal by sellers and Amazon representatives. Although there will be a high number of reads and writes, customers expect subsecond response times regarding search results. This can be achieved by maintaining separate write and search databases.

Hence, this use case is ideal for the CQRS pattern. Here, when a customer searches, data is fetched from the search database, and when the customer orders or adds something to the cart, it is written to the write database. The information available in the write database will be streamed in real time to a search database such as Elasticsearch or AWS OpenSearch. So, users who are searching for products and discounts should get the search results in a fraction of a second.

The strangler fig pattern

The strangler fig pattern derives its name from a species of tropical fig plants that grow around their host trees, slowly strangling the host tree so that it dies. This pattern was first proposed by Martin Fowler. Although the basic pattern may be implemented in different ways, the streaming pipeline gives us an indigenous way to use this pattern. To understand this pattern, let's look at an example.

Suppose there is a monolithic application that consists of three modules – A, B, and C. A, B, and C read and write data to the database. Initially, the architecture looked as follows:

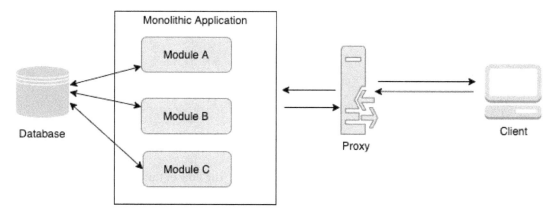

Figure 7.12 – Initial state of the monolithic application

As we can see, all the modules have double-ended arrows, denoting both reads and writes are happening. Now, using the strangler fig pattern, we can convert this monolithic legacy application into a microservices-based application by slowly migrating the individual modules as separate microservices – one at a time. The following diagram shows that **Module A** is being moved to the microservices pattern from the monolithic app:

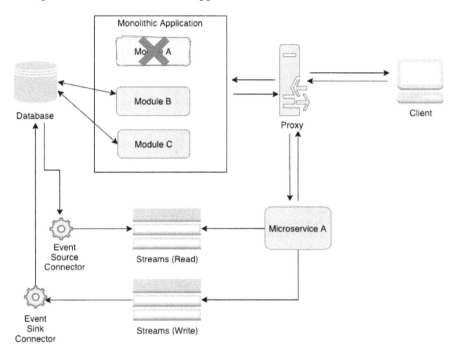

Figure 7.13 – Module A replaced with Microservice A

As we can see, **Microservice A** (which has successfully replaced **Module A**) reads and writes data to an event stream. This event stream is, in turn, connected to the database using an event source or sink connector. Slowly, the monolithic application will be strangled, and the final transformed architecture will look as follows:

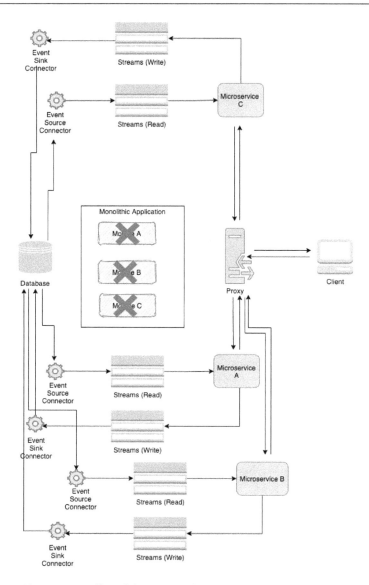

Figure 7.14 – All modules migrated using the strangler fig pattern

As we can see, all the modules have been migrated from the monolithic application to the federated microservice pattern, allowing the monolithic application to retire.

The log stream analytics pattern

In this pattern, we will learn how logs collected across various apps, web portals, backend services, and IoT devices are used for analytics and monitoring. The following diagram shows a typical log streaming pattern used to facilitate log analytics and monitoring:

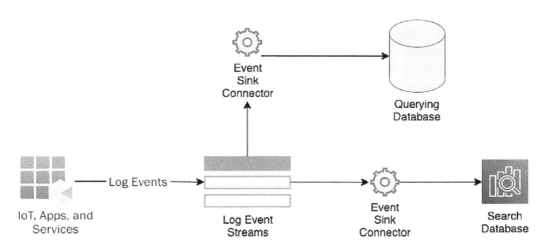

Figure 7.15 – The log stream analytics pattern

Let's learn how this pattern works:

1. As evident from the preceding diagram, all log events from various IoT devices, apps, web portals, and services are streamed into an event stream.

2. Then, using the event sink connector, the events are sent to both a search database and a querying database.

A search database can be a search engine such as Elasticsearch, AWS OpenSearch, Splunk, or Apache Solar. This database facilitates quick searches with complex query patterns. It also enables visualization and analytics using the capabilities of the search engine. The query database is either an MPP database such as Redshift or Snowflake or a query engine such as Athena. A query engine allows users to run SQL queries on top of ObjectStores such as S3 objects.

The following diagram shows a sample implementation of this kind of pattern in AWS:

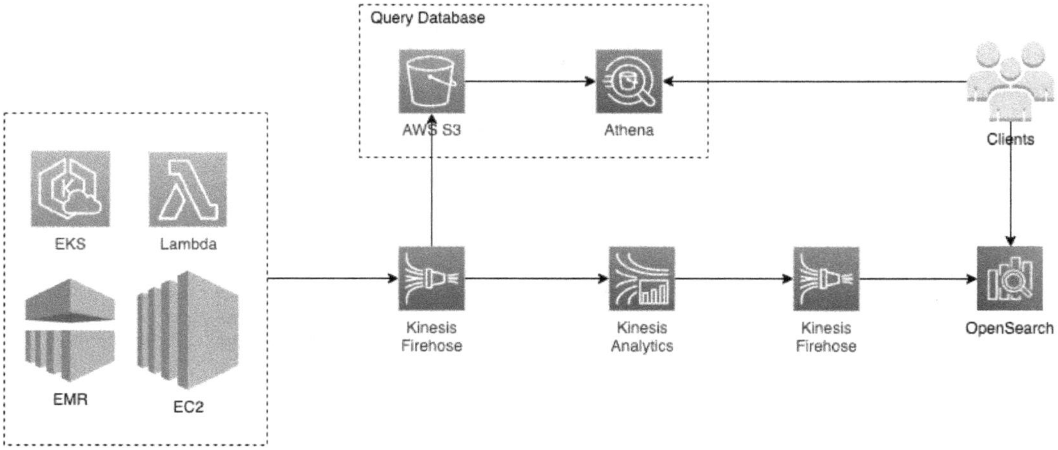

Figure 7.16 – Example of a log analytics pattern in AWS

Here, log events from various AWS services such as EC2, ECR, EKS, and others are streamed to a Kinesis topic using Kinesis Firehose. Kinesis Analytics transformation is done and, using Kinesis Firehose, streamed to AWS OpenSearch for search and analytics purposes. On the other hand, data gets streamed into S3 from the first Kinesis Firehose. Athena tables are created on top of the S3 objects. Athena then provides an easy-to-use query interface for batch-based analytic queries to be performed on the log data.

In this section, we learned about various streaming patterns that are popular and can be used to solve common data engineering problems. We also looked at a few examples and learned when these patterns should be used. Next, we will investigate a few popular patterns that are a mix of both batch and stream processing. These are known as hybrid data processing patterns.

Hybrid data processing patterns

In this section, we will discuss two very famous patterns that support both batch and real-time processing. Since these patterns support both batch processing and stream processing, they are categorized as hybrid patterns. Let's take a look at the most popular hybrid architectural patterns.

The Lambda architecture

First, let's understand the need for Lambda architecture. In distributed computing, the CAP theorem states that any distributed data can guarantee only two out of the three features of the data – that is, consistency, availability, and partition tolerance. However, Nathan Marz proposed a new pattern in 2011 that made it possible to have all three characteristics present in a distributed data store. This pattern is called the Lambda pattern. The Lambda architecture consists of three layers, as follows:

- **Batch layer**: This layer is responsible for batch processing

- **Speed layer**: This layer is responsible for real-time processing

- **Serving layer**: This layer serves as the unified serving layer where querying can be done by downstream applications

The following diagram shows an overview of the Lambda architecture:

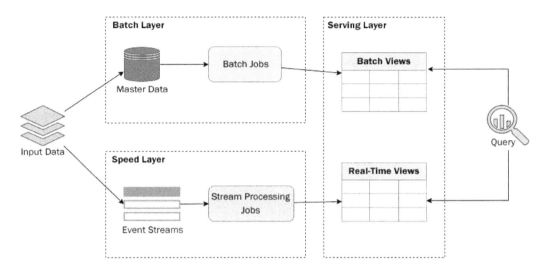

Figure 7.17 – The Lambda architecture

In the Lambda architecture, the input data or the source data is written to the master data store present in the batch layer, as well as the event streams present in the speed layer. The master data store may be a relational or NoSQL database or a filesystem such as HDFS. Batch processing jobs run on top of this data store to do any data processing, as well as to load the data into batch views present in the serving layer. Events written in the event stream are picked up, processed, and loaded into the real-time view by stream processing jobs (as shown in *Figure 7.17*). Queries can be made separately to query batch views and real-time views, or they can be queried simultaneously on both views to view the results. Batch views are mainly for historical data, while real-time views are for Delta data.

Although it solves the problem of eventual consistency as queries can combine data from both real-time views and batch-based views, it comes with a few shortcomings. One of the major shortcomings is that we must maintain two different workflows – one for the batch layer and another for the speed layer. Since, in a lot of scenarios, the technology to implement a streaming application is quite different from a batch-based application, we must maintain two different source codes. Also, debugging and monitoring for both batch and stream processing systems becomes an overhead. We will discuss how to overcome these challenges in the next pattern.

The Kappa architecture

One of the reasons the Lambda architecture is widely accepted is because it can overcome the limitation of the CAP theorem and enable more use of stream processing across the industry. Before the Lambda architecture, businesses were skeptical to use stream processing as they feared losing messages during real-time processing. However, this assumption is not true with modern distributed streaming platforms such as Kafka and Pulsar. Let's take a look at how the Kappa architecture provides a simpler alternative to the Lambda architecture. The following diagram depicts the Kappa architecture:

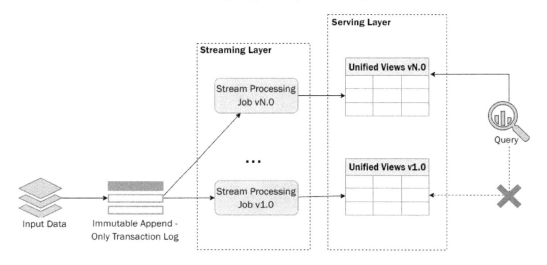

Figure 7.18 – The Kappa architecture

In the Kappa architecture, the idea is not to use two different flows – one for batch and one for streaming. Instead, it proposes all processing should be done using a stream processing engine. This means that both batch-based workloads and stream-based workloads can be handled by a single pipeline. Here, input data or source data is written in a special event stream. This event stream is an immutable append-only transaction log. Since it is an append-only log, it has fast writing capabilities. To read data, we can read from the offset where the data read stopped earlier. In addition to this last read offset, it should support replayability, which means we can read from the first message as well.

Since we are talking about distributed computing, this transaction log will be partitioned, which will improve the read and write performance as well. Stream processing jobs read the events, process them, and write the output to unified views (containing both batch and real-time data). One question that comes to mind is, *How does this kind of flow support high-volume batch loads?* Huge volumes of data are also sent to the transaction log, which is then picked up by stream processing jobs. The output is stored in the view present in the serving layer. To process such a high-volume event stream, we need to do more parallelism by increasing the number of partitions in the event stream.

This event stream log should have retention capabilities. Also, consumers should be able to replay the stream using either the event time or event offset. Each event has an offset and an event timestamp.

The replayability feature allows consumers to re-read already fetched data by setting the event offset or event timestamp to an older value.

So far, we have discussed commonly used batch-based, real-time, and hybrid architectural patterns. In the penultimate section, we will quickly look at a few common serverless patterns.

Serverless patterns for data ingestion

We will start by answering the question, *What is serverless computing?* Serverless computing is a cloud execution model in which a cloud provider takes care of allocating resources such as storage and compute based on demand while taking care of the servers on behalf of customers. Serverless computing removes the burden of maintaining and managing servers and resources associated with it. Here, the customers of serverless computing don't care how and where the jobs or applications are running. They just focus on the business logic and let the cloud provider take care of managing the resources for running and executing that code. A few examples of serverless computing are as follows:

- **AWS Lambda Function or Azure Function**: This is used to run any application or service
- **AWS Glue**: This is used to run big data-based ETL jobs
- **AWS Kinesis**: This is a serverless event streaming and analytics platform

Although there are many useful serverless patterns, in this section, we will discuss the two most relevant patterns that can help us architect data engineering solutions. The following are the serverless patterns that we will be discussing:

- **The event-driven trigger pattern**: This is a very common pattern that's used in cloud architectures. In this pattern, upon creating or updating any file in object storage such as an S3 bucket, a serverless function gets triggered. The following diagram provides an overview of this pattern:

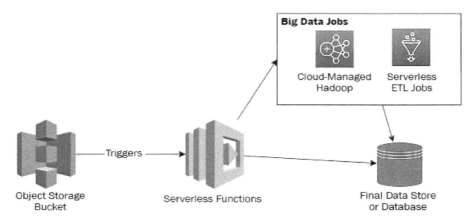

Figure 7.19 – The event-driven trigger pattern

In this pattern, any change to an object, such as it being created or deleted, in object storage can trigger a serverless function. This serverless function may either directly process the data or be used to trigger a big data job. Serverless functions such as AWS Lambda and Azure Function can set triggers that can trigger them. For example, a Lambda function can be configured to have an S3 trigger from Bucket1 for any new object being created or updated in Bucket1. The triggered Lambda function can, in turn, trigger an EMR job or a serverless Glue job, which transforms and processes the necessary data and writes the final output to a data store. Alternatively, the Lambda function can do some data processing and store the output result in the final data store. The final data store can be a SQL database, NoSQL database, MPP database, or object storage such as AWS S3.

A real-world scenario for using this pattern and its solution was explained in detail in *Chapter 5, Architecting a Batch Processing Pipeline*.

- **The serverless real-time pattern**: This is an oversimplistic serverless pattern that is quite popular for data ingestion in the cloud. An overview of this pattern can be seen in the following diagram:

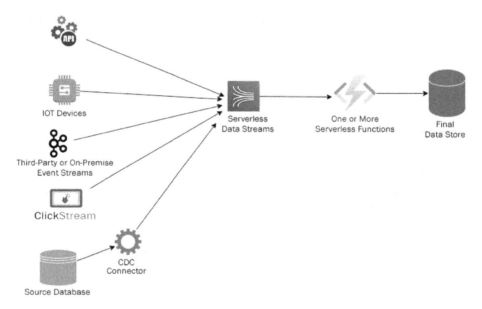

Figure 7.20 – The serverless real-time pattern

In the serverless real-time pattern, event or data streaming, as well as data processing, happens using serverless services in the cloud. Events, logs, and messages from different source systems publish the events to a serverless data streaming platform such as AWS Kinesis. The data stream triggers one or a series of serverless functions chained one after another to do the data processing on the fly. Once the data has been processed, it is written back to a final data store. The final data store can be SQL, NoSQL, an MPP database, object storage, or a search engine.

A real-world example where this pattern may be used is in a real-time fraud detection system for credit card usage. The following diagram depicts a sample solution of fraud detection in AWS using this pattern:

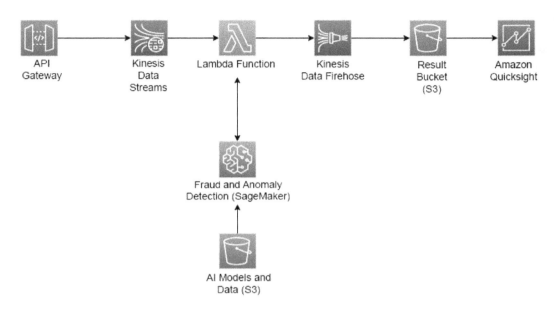

Figure 7.21 – A sample implementation of the serverless real-time pattern in AWS

Here, API Gateway streams real-time credit card transactions directly into Kinesis Data Streams (a serverless data streaming platform). The transaction events written in Kinesis Data Streams trigger the Lambda function to perform fraud and anomaly detection on the event. The Lambda function makes use of AWS SageMaker, which, in turn, uses the already stored data science models stored in S3 to determine fraud and anomalies in the transaction. The output is then passed to Kinesis Data Firehose, which captures the result from the Lambda function and stores it in an S3 bucket. This S3 bucket contains the results in real time. We can use a service such as Amazon QuickSight to visualize the results and take any action if required.

With that, we have discussed what serverless computing is and discussed two highly used patterns for serverless computing for data ingestion. Now, let's summarize what we learned in this chapter.

Summary

In this chapter, we started by discussing various popular batch processing patterns. We covered five commonly used patterns to solve batch processing problems. We also looked at examples of those patterns and real-world scenarios where such patterns are used. Then, we looked at five popular patterns available to architect stream processing pipelines and how they are used to solve real-world problems in data engineering. Next, we learned about the Lambda and Kappa architectures and how they are useful for both batch and stream processing. Finally, we learned what serverless architecture is and looked at two popular serverless architectures that are used to solve many data engineering problems in the cloud.

At this point, we know how to implement batch and streaming solutions, as well as have a fair idea of different data engineering patterns that are commonly used across the industry. Now, it is time to put some amount of security and data governance into our solutions. In the next chapter, we will discuss various data governance techniques and tools. We will also cover how and why data security needs to be applied to data engineering solutions.

8

Enabling Data Security and Governance

In the preceding chapters, we learned how to evaluate requirements and analyze and apply various architectural patterns to solve both real-time and batch-based problems. We learned how to choose the optimal technical stack and develop, deploy, and execute the proposed solution. We also discussed various popular architectural patterns for data ingestion. However, any discussion about data architecture is incomplete without mentioning data governance and data security. In this chapter, we will focus on understanding and applying data governance and security in the data layer.

In this chapter, we will first discuss what data governance is, and why it is so important. We will also briefly discuss a few open source data governance tools that are available on the market. Then, we will practically demonstrate a data governance implementation by adding a data governance layer to a data ingestion pipeline. The data ingestion pipeline ID will be developed using Apache NiFi and data governance will be achieved using DataHub. Then, we will discuss the need for data security and the types of solutions that help enable it. Finally, we will discuss a few open source tools available to enable data security.

By the end of the chapter, you will know the definition of and need for a data governance framework. You will also know when data governance is required and all about the data governance framework by the **Data Governance Institute** (**DGI**). In addition, you will know how to implement practical data governance using DataHub. Finally, you will know about various solutions and tools available to enable data security.

In this chapter, we're going to cover the following main topics:

- Introducing data governance – what and why
- Practical data governance using DataHub and NiFi
- Understanding the need for data security
- Solutions and tools available for data security

Technical requirements

For this chapter, you will need the following:

- OpenJDK 1.11 installed on your local machine
- Docker installed on your local machine
- An AWS account
- NiFi 1.12.0 installed on your local machine
- DataHub installed on your local machine
- Prior knowledge of YAML is preferred

The code for this chapter can be downloaded from this book's GitHub repository: `https://github.com/PacktPublishing/Scalable-Data-Architecture-with-Java/tree/main/Chapter08`.

Introducing data governance – what and why

First, let's try to understand what data governance is and what it does. In layman's terms, **data governance** is the process of assigning proper authority and taking decisions regarding data-related matters. According to the DGI, data governance is defined as "*a system of decision rights and accountabilities for information-related processes, executed according to agreed-upon models that describe who can take what actions with what information, and when, under what circumstances, using what methods.*"

As evident from this definition, it is a practice of creating definitive strategies, policies, and rules that define who can make what decisions or perform actions related to data. It also lays out guidelines about how to make decisions related to data.

Data governance considers the following aspects:

- Rules
- Enterprise-level organizations
- Decision rights and procedures
- Accountability
- Monitoring and controlling data

Now that we have a basic understanding of data governance, let's explore what scenarios data governance is recommended in.

When to consider data governance

A formal data governance framework should be considered in any of the following scenarios:

- The data in the organization has grown so much and become so complex that traditional data management tools are unable to solve cross-functional data needs.

- The horizontally focused business units and teams need cross-functional data generated or maintained by other focused groups. In this case, an enterprise-wide solution (instead of siloed) for data availability and management is required. For example, for B2B sales of Visa, the marketing and accounting departments maintain data in silos. However, the sales department needs both these silos' data to create accurate sales predictions. So, the data across these different departments should be stored in an enterprise-wide central repository instead of silos. Hence, this enterprise-wide data requires data governance for proper access, use, and management.

- Compliance, legal, and contractual obligations can also call for formal data governance. For example, the **Health Insurance Portability and Accountability Act (HIPAA)** enforces that all **protected health information (PHI)** data should be well governed and protected from theft or unauthorized access.

So far, we've discussed what data governance is and when we should have a formal data governance framework. Now we will learn about the DGI data governance framework.

The DGI data governance framework

The DGI data governance framework is a logical structure for implementing proper data governance in an organization to enable better data-related decisions to be made. The following diagram depicts this data governance framework:

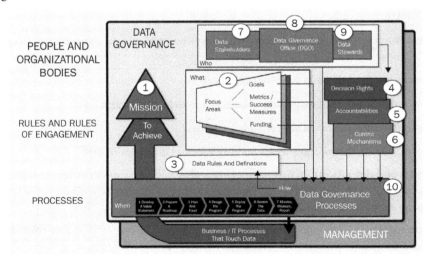

Figure 8.1 – Data governance framework

In this section, we will discuss the various components of the DGI data governance framework. Let's take a look at each of the components highlighted in the preceding diagram:

1. **Mission**: The mission of the DGI framework can be attributed to three primary responsibilities:

    A. Define rules

    B. Execute and implement the rules by providing ongoing protection and services to data stakeholders

    C. Deal with scenarios that arise due to non-compliance

    This form of mission is very similar to political governance. In the political governance model, the government makes the rules, the executive branch implements the rules, and the judiciary deals with non-compliance or rule breakers. Just like political governance, here, one set of data stakeholders defines the rules. Another set of stakeholders ensures that the rules are followed. Finally, a third set of stakeholders makes decisions related to non-compliance. Optionally, the organization can develop a vision statement around the mission, which may be used to inspire data stakeholders to envision possibilities and set data-related goals.

2. **Focus areas**: Mission and vision lead us to the primary focus areas. We have two primary focus areas. They are as follows:

    - Goals should be **SMART** – that is, **Specific, Measurable, Actionable, Relevant, and Timely**. We should remember the principle of the four P's while deciding on the goals we want to pursue – that is, programs, projects, professional disciplines, and people. We must ask how these efforts can help our organization in terms of revenue, cost, complexity, and ensuring survival (that is, security, compliance, and privacy).

    - Metrics should also be SMART. Everyone in data governance should know how to quantify and measure success.

    This discussion leads to the question of where we can fund our data governance program. To do so, we must ask the following questions:

    - How can we fund the data governance office?

    - How can we fund the data architects/analysts to define rules and data?

    - How can we fund the data stewardship activities?

    Focus areas are very important when planning formal data governance.

3. **Data rules and definitions**: In this component, rules, policies, standards, and compliances are set around data. Typical activities may include creating new rules, exploring existing rules, and addressing gaps and overlaps.

4. **Decision rights**: This component implores the question, *Who can make a data-related decision, when, and using what process?* The data governance program allows us to store decision rights as metadata for data-related decisions.

5. **Accountability**: This component creates accountability for the implementation of a rule or decision once it is made. The data governance program may be expected to integrate the accountabilities in a day-to-day **software development life cycle (SDLC)**.

6. **Controls**: Data is the new gold, so there is a huge security risk involved if there are sensitive data breaches. How do we ensure that such risks are mitigated and handled properly? They can be handled using controls. Controls can be preventive or reactive. The data governance team is usually tasked with making recommendations for creating these controls at different levels of the control stack (networking/OS/database/application). Sometimes, the data governance team is also asked to modify the existing controls to ensure better data governance.

7. **Data stakeholders**: Data stakeholders are individuals or groups who either could affect or are affected by the data. Since they have a direct correlation with the data, they are consulted while taking data decisions. Again, based on the scenario, they may want to be involved in some data-related decisions, should be consulted before decisions are finalized, or be informed once decisions are made.

8. **Data governance office**: This facilitates, supports, and runs the data governance program and data stewardship activities. It is the centralized governing body and usually consists of data architects, data analysts, and those working on creating the metadata. It collects and aligns policies, rules, and standards from different stakeholders across the organization and comes up with organizational-level rules and standards. It is responsible for providing centralized data governance-related communication. It is also responsible for collecting the data governance metrics and publishing reports and success measures to all the data stakeholders.

9. **Data stewards**: They are part of the data stewardship council, which is responsible for making data-related decisions such as setting policies, specifying standards, or providing recommendations to the DGO. Based on the size and complexity of the organization, the data stewardship council(s) may have a hierarchy. In data quality-focused governance projects, there can be an optional data quality steward.

10. **Data processes**: These are the methods that are used to govern data. The processes should be documented, standardized, and repeatable. They are designed to follow regulatory and compliance requirements for data management, privacy, and security.

All these components work in tandem and create a feedback loop, which ensures data governance is continuously improving and up to date.

In this section, we learned about what data governance is, when it should be implemented, and the DGI framework. In the next section, we will provide step-by-step guidelines for implementing data governance.

Practical data governance using DataHub and NiFi

In this section, we will discuss a tool called DataHub and how different data stakeholders and stewards can make use of it to enable better data governance. But first, we will understand the use case and what we are trying to achieve.

In this section, we will build a data governance capability around a data ingestion pipeline. This data ingestion pipeline will fetch any new objects from an S3 location, enrich them, and store the data in a MySQL table. In this particular use case, we are getting telephone recharge or top-up events in an S3 bucket from various sources such as mobile or the web. We are enriching this data and storing it in a MySQL database using an Apache NiFi pipeline.

Apache NiFi is a powerful and reliable drag-and-drop visual tool that allows you to easily process and distribute data. It creates directed graphs to create a workflow or a data pipeline. It consists of the following high-level components so that you can create reliable data routing and transformation capabilities:

- **FlowFile**: Each data record is serialized and processed as a FlowFile object in the NiFi pipeline. A FlowFile object consists of flowfile content and attribute denoting data content and metadata respectively.

- **Processor**: This is one of the basic units of NiFi. This component is primarily responsible for processing the data. It takes a FlowFile as input, processes it, and generates a new one. There are around 300 inbuilt processors. NiFi allows you to develop and deploy additional custom processors using Java.

- **Queue**: NiFi follows a staged event-driven architecture. This means that communication between processors is asynchronous. So, there needs to be a message bus that will hold a FlowFile generated by one processor, before it is picked by the next processor. This message bus is called the NiFi queue. Additionally, it supports setting up backpressure thresholds, load balancing, and prioritization policies.

- **Controller service**: This allows you to share functionality and state across JVM cleanly and consistently. It is responsible for things such as creating and maintaining database connection pools or distributed caches.

- **Processor groups**: When a data flow becomes complex, it makes more sense to group a set of components, such as processors and queues, into an encapsulation called a processor group. A complex pipeline may have multiple processor groups connected. Each processor group, in turn, has a data flow inside it.

- **Ports**: Processor groups can be connected using ports. To get the input from an external processor group, input ports are used. To send a processed FlowFile out of the processor group, output ports are used.

Now, let's build the NiFi pipeline for our use case. Our source is an S3 bucket called `chapter8input`, while our output is a MySQL cluster. We will be receiving S3 objects in the `chapter8input` folder from different sources. Each S3 object will be in JSON format and will contain one phone recharge or top-up (bundle) event. Our data sink is a MySQL table called `bundle_events`. The **Data Definition Language** (**DDL**) of this table is as follows:

```
CREATE TABLE `dbmaster`.`bundle_events` (
  `customerid` INT NOT NULL,
  `bundleid` INT NOT NULL,
  `timestamp` VARCHAR(45) NOT NULL,
  `source` VARCHAR(45) NULL,
  PRIMARY KEY (`customerid`, `bundleid`, `timestamp`));
```

Now, the NiFi pipeline polls the S3 bucket and checks for any change events, such as a JSON file being created or updated. Once a file is uploaded to the S3 bucket, NiFi should fetch the file, enrich it with the source type using the S3 object name, and then write the enriched data to the MySQL `bundle_events` table.

If you don't have Apache NiFi installed in your system, download and install Apache NiFi-1.12.0. You can download, install, and follow the startup instructions at `https://github.com/apache/nifi/blob/rel/nifi-1.12.0/nifi-docs/src/main/asciidoc/getting-started.adoc#downloading-and-installing-nifi` to do so. Alternatively, you can spin up a NiFi cluster/node on an AWS EC2 instance.

Creating the NiFi pipeline

Next, we will discuss how to build the NiFi pipeline. The NiFi pipeline that we will build is shown in the following screenshot:

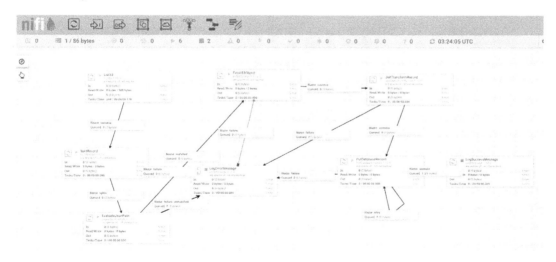

Figure 8.2 – NiFi pipeline to read data from S3 and write to MySQL

Let's try to understand the overall NiFi pipeline and how we are building it:

1. First, we have used a **ListS3** processor to capture whether any S3 objects have been inserted or updated in the configured S3 bucket. It lists all the change events.

2. Then, using the **SplitRecord** processor, we split the records into individual events.

3. Using the **EvaluateJsonPath** processor, we create the sourceKey attribute for the FlowFile.

4. Then, we use **FetchS3Object** to fetch the S3 object. The **FetchS3Oject** processor is responsible for reading the actual S3 object. If **FetchS3Object** successfully reads the file, it sends the S3 object content as a FlowFile to the **JoltTransformRecord** processor.

5. **JoltTransformRecord** is used to enrich the data before the enriched data is sent to be written to MySQL using the **PutDatabaseRecord** processor.

6. The success of the **PutDatabaseRecord** processor is sent to the **LogSuccessMessage** processor. As shown in the preceding screenshot, all FlowFiles in failure scenarios are sent to the **LogErrorMessage** processor.

Now, let's configure each of the NiFi processors present in the NiFi pipeline (shown in *Figure 8.2*):

- **ListS3 processor**: The configuration of the **ListS3** processor is shown in the following screenshot:

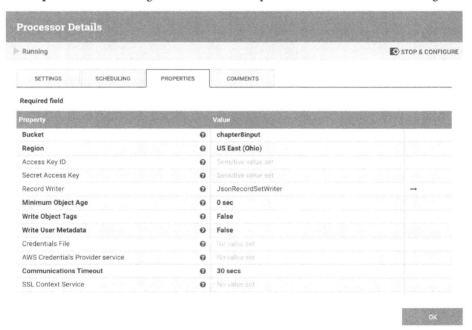

Figure 8.3 – Configuring the ListS3 processor

As we can see, the **ListS3** processor is configured to poll and listen for changes in the **chapter8input** S3 bucket. Apart from the bucket name, we must configure the **Region**, **Access Key ID**, and **Secret Access Key** details for the NiFi instance to connect to the AWS S3 bucket. Finally, we have configured the Record Writer property, which is set as a controller service of the **JsonRecordSetWriter** type.

- **SplitRecord processor**: Next, we will configure the **SplitRecord** processor, as shown in the following screenshot:

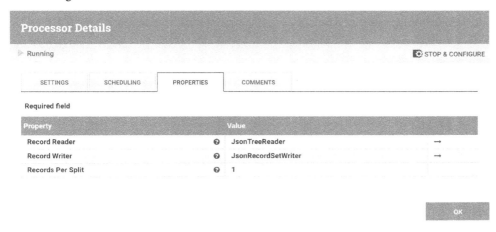

Figure 8.4 – Configuring the SplitRecord processor

SplitRecord is responsible for splitting a single FlowFile containing multiple write events in S3 into individual events. Now, each event is metadata for one **S3Object**.

- **EvaluateJsonPath processor**: We use the **EvaluateJsonPath** processor to extract the value of the **key** column from the FlowFile content and add it as an attribute to the FlowFile attributes. The configuration of the **EvaluateJsonPath** processor is shown in the following screenshot:

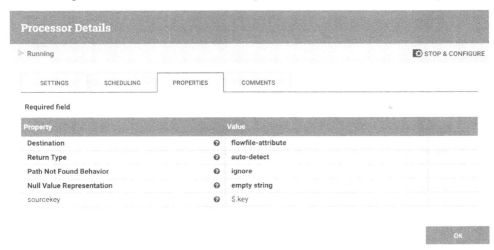

Figure 8.5 – Configuring the EvaluateJsonPath processor

Here, we have configured the **Destination** property with the `flowfile-attribute` value. This indicates that a new key-value pair will be added to the FlowFile attributes. The attribute to be added should be provided as a dynamic property, where the name of the property will be the attribute key and the value of the property will be the attribute value. Here, we have added a property named **sourcekey**. The value of this property is the NiFi expression `$.key`. This expression fetches the value of the `key` field from the FlowFile content (which is a JSON).

- **FetchS3Object processor**: The following screenshot shows the configuration of the **FetchS3Object** processor:

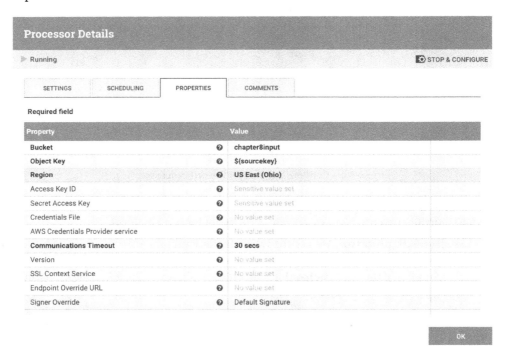

Figure 8.6 – Configuring the FetchS3Object processor

As we can see, the **FetchS3Object** configuration uses the newly added **sourcekey** attribute. `${sourcekey}` is a NiFi expression that gets the value of the **sourcekey** attribute. Apart from this, the S3-related properties such as **Bucket**, **Access Key ID**, and **Secret Access Key** need to be set in this processor. The output of this processor is the content of our **S3Object** (which is in JSON format).

- **JoltTransformRecord processor**: The **JoltTransform** processor is used to add a new key-value pair to the JSON. The configuration of the **JoltTransformRecord** processor is as follows:

Figure 8.7 – Configuring the JoltTransformRecord processor

The **JoltTransformRecord** processor uses Apache Jolt to do JSON transformations. Jolt takes a JSON called **Jolt Specification** to transform one JSON into another. In a **Jolt Specification**, multiple types of operations can be performed, such as `shift`, `chain`, and `default`. For more information about Apache Jolt, go to `https://github.com/bazaarvoice/jolt#jolt`.

Here, we will discuss the Jolt transformation we are using to add a key-value pair to the JSON FlowFile content. We add the `source` key using the **Jolt Specification** property, as follows:

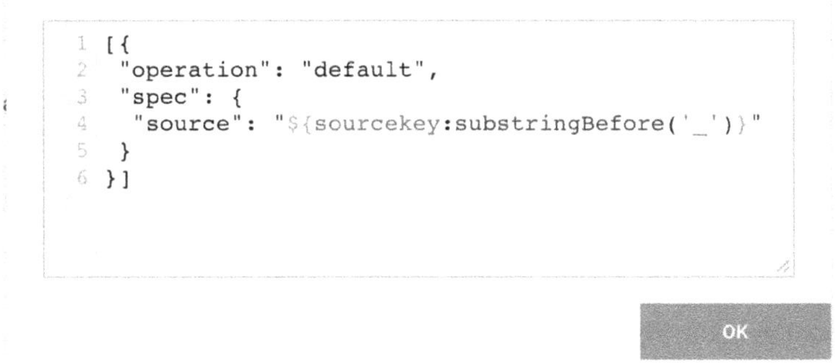

```
1  [{
2    "operation": "default",
3    "spec": {
4      "source": "${sourcekey:substringBefore('_')}"
5    }
6  }]
```

Figure 8.8 – Using the Jolt Specification property to add the source key

As you can see, by setting `operation` to `default` in **Jolt Specification**, you can add a new key-value pair to an input JSON. Here, we are adding a `source` key with the dynamic value calculated using the NiFi expression language. `${sourcekey:substringBefore('_')}` is a NiFi expression. This expression returns a substring of the `FlowFile` attribute's **sourcekey**, from the beginning of the string till the character before the first occurrence of an underscore (`'_'`).

- **PutDatabaseRecord processor**: Once this JSON is enriched with the newly added key-value pair, the record is written to the MySQL database using the **PutDatabaseRecord** processor. The following screenshot shows the configuration of the **PutDatabaseRecord** processor:

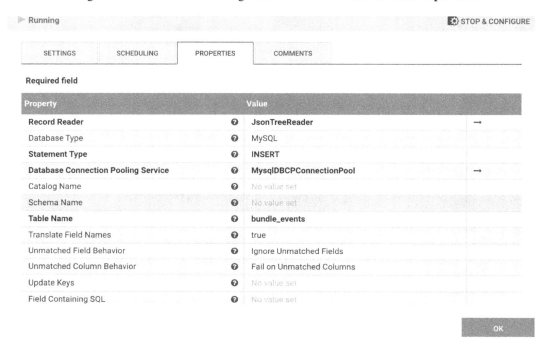

Figure 8.9 – Configuring the PutDatabaseRecord processor

As evident from this configuration, we need to configure two controller services – **Record Reader** and **Database Connection Pooling Service**. We configure a simple **JsonTreeReader** service for **Record Reader** and a **DBCPConnectionPool** service for **Database Connection Pooling Service**. Apart from that, we also need to set **Database Type** to MySQL, **Statement Type** to INSERT, and **Table Name** to bundle_events. Now, let's see how the **DBCPConnectionPool** service, called **MysqlDBCPConnectionPool**, is configured. The following screenshot shows the configuration of the **MysqlDBCPConnectionPool** service:

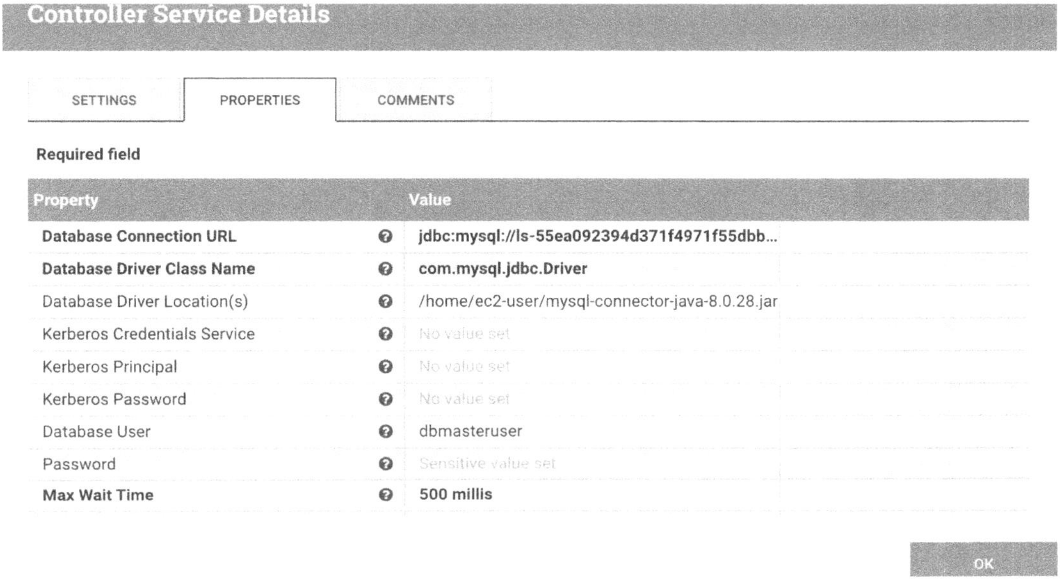

Figure 8.10 – Configuring the DBCPConnectionPool service

The **MysqlDBCPConnectionPool** controller service is used to create a JDBC connection pool. To do so, you must configure the **Database Connection URL** (the JDBC URL), **Database Driver Class Name**, and **Database Driver Location(s)** properties, as shown in the preceding screenshot.

With that, we have built the entire NiFi pipeline to extract data from S3, enrich it, and write it to a MySQL table. You can check out the entire NiFi pipeline by going to `https://github.com/PacktPublishing/Scalable-Data-Architecture-with-Java/blob/main/Chapter08/sourcecode/nifi_s3ToMysql.xml`.

Setting up DataHub

Now, we will add a data governance layer to our solution. Although many data governance tools are available, most of them are paid tools. There are several data governance tools available for pay-as-you-go models in the cloud, such as AWS Glue Catalog. However, since our solution will be running on-premises, we will choose a platform-agnostic open source tool. DataHub, developed by LinkedIn, is one such open source tool that comes with a decent set of features. We will use it in this chapter to explain data governance practically. In this section, we will learn how to configure DataHub. In this pipeline, S3 is the source, MySQL is the target, and NiFi is the processing engine. To create data governance, we need to connect to all these systems and extract metadata from them. DataHub does this for us.

Before we begin connecting these components (of the data pipeline) to DataHub, we need to download and install DataHub in our local Docker environment. The detailed installation instructions can be found at `https://datahubproject.io/docs/quickstart`.

In the next section, we will learn how to connect different data sources to DataHub.

Adding an ingestion source to DataHub

To connect a data source or pipeline component to DataHub, we must go to the **Ingestion** tab from the top-right menu bar, as shown in the following screenshot:

Figure 8.11 – Connecting a new source for metadata ingestion in DataHub

Then, we must click the **Create new source** button to create a new connection to a data source. On clicking this button, we get the following popup:

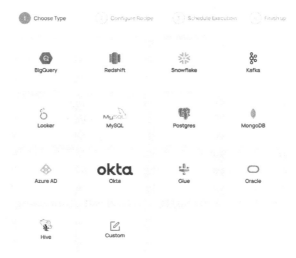

Figure 8.12 – Choosing the type of data source to create a new connection

This popup is a multipage wizard where you can create a new data source connection. First, as shown in the preceding screenshot, you are asked to choose a data source type. For example, if the data is in a MySQL database, then we would select MySQL from the various thumbnails present in the wizard. For sources, which are not listed, such as NiFi and S3, we can select the **Custom** thumbnail. Once you've chosen a type, click **Next**; you will be taken to the second step called **Configure Recipe**, as shown in the following screenshot:

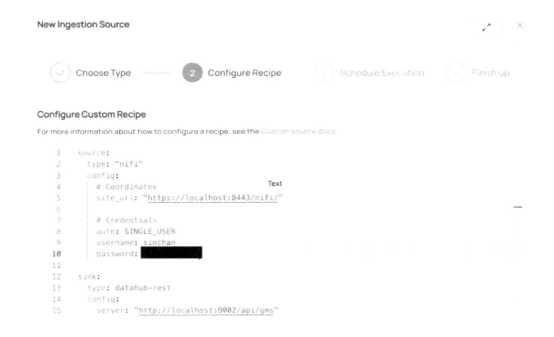

Figure 8.13 – Configuring a recipe for the data source

In the **Configure Recipe** step, we have to add some YAML code in the space provided. Irrespective of the data source, the YAML has two top-level elements – `source` and `sink`. Again, both `source` and `sink` contain two elements – `type` and `config`. Here, `type` denotes the type of source or sink. For example, in the preceding screenshot, we are configuring NiFi, so the source type is `nifi`. If we had selected MySQL as the type of connection in the previous step, the source's `type` in this step would be `mysql`.

There are primarily three types of sinks that are currently available in DataHub, as follows:

- **Console**: The metadata is sent to `stdout`
- **DataHub**: The metadata is sent to DataHub via the GMS REST API
- **File**: The metadata is sent to a configured file

Here, we will be using DataHub as the sink since this allows us to do governance and monitoring using DataHub's capabilities. In this case, our type of sink will be datahub-rest. We also need to specify the GMS REST API's HTTP address as the base path. Since our DataHub installation is using Docker, we will use localhost:9002 as the server IP address (as shown in *Figure 8.13*).

Once everything has been added, click **Next** to go to the **Schedule Execution** step, as shown in the following screenshot:

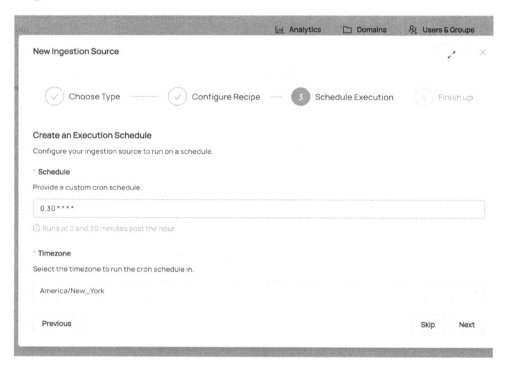

Figure 8.14 – Create an execution schedule

In this step, we set an execution schedule for DataHub to fetch metadata from the source. In this case, we configured a CRON expression, 0,30 * * * *, which means it will run every 30 minutes.

Click **Next** to go to the last step, as shown in the following screenshot:

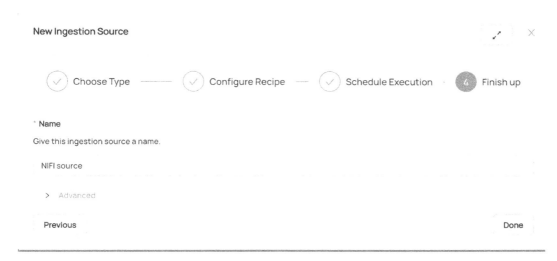

Figure 8.15 – The final step of the New Ingestion Source wizard

In the final step of the wizard, we need to enter the name of the ingestion source. This name can be any arbitrary name that helps uniquely identify that source. Finally, we must click **Done** to add the new data source. The following screenshot shows the added sources:

Manage Ingestion

Create, schedule, and run DataHub ingestion pipelines

Sources   Secrets

+ Create new source   ◯ Refresh

	Type	Name	Schedule	Execution Count	Last Execution	Last Status
+	MySQL	mysql	0,30 * * * *	42	6/5/2022 at 5:30:04 PM	Running
+	Custom	NIFI source	0,30 * * * *	62	6/5/2022 at 5:30:04 PM	⊙ Failed
+	Custom	S3input source	0,30 * * * *	36	6/5/2022 at 5:30:04 PM	Running

Figure 8.16 – Sources added at a glance

As shown in the preceding screenshot, we can monitor how many times a job has run to fetch metadata from each source and whether the last run was successful or not. All the YAML files that have been used for our use case can be found on GitHub at https://github.com/PacktPublishing/Scalable-Data-Architecture-with-Java/tree/main/Chapter08.

In the next section, we will discuss how we can perform different data governance tasks using this tool.

Governance activities

As discussed in the preceding section, once a data source or pipeline is connected with DataHub, it provides a lot of tools to enable a data governance model around those sources. The following are a few capabilities that we will explore in this chapter:

- **Adding a domain**: A domain in DataHub can be any logical grouping; it might be organization-specific. This helps us analyze the domain-wise resource utilization and other statistics. For example, an organization can use business unit names as domains. In our particular scenario, **DataServices** is a business unit in the organization, and we create a domain in the name of the business unit. To create a new domain, we can navigate to **Manage | Domains** and click the **New Domain** button. Once this button is clicked, a popup dialog will open, as shown in the following screenshot:

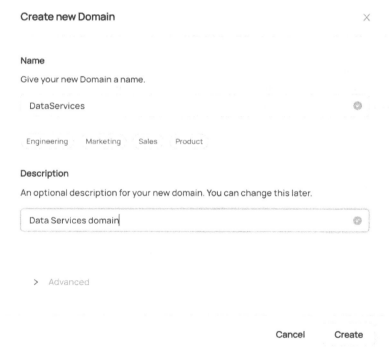

Figure 8.17 – Creating a new domain

As we can see, we must provide a name and description for a domain. You cannot create two domains with the same name.

- **Adding a user group**: We can manage users and groups by going to **Settings** and selecting the **Groups** tab. Upon clicking the **Create group** button, a dialog similar to the following will appear:

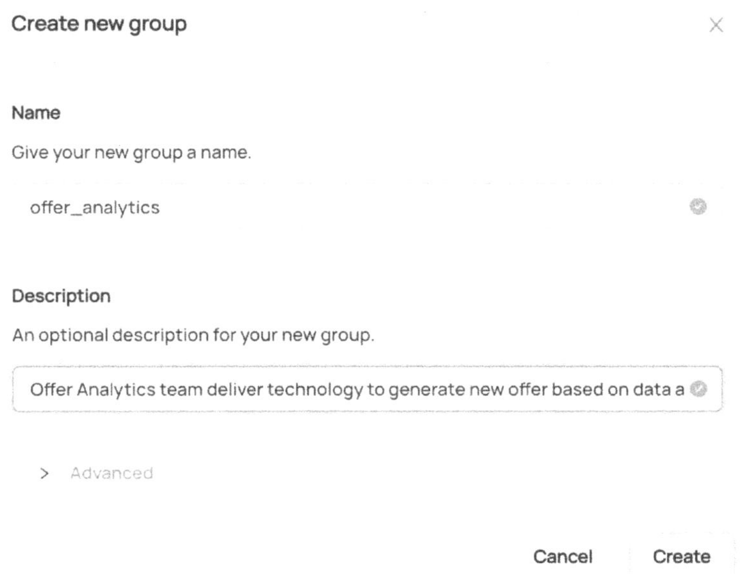

Create new group
Name
Give your new group a name.
offer_analytics
Description
An optional description for your new group.
Offer Analytics team deliver technology to generate new offer based on data a
> Advanced
Cancel
Create

Figure 8.18 – Adding a new user group

As we can see, we must provide a group name and an optional description. User groups help make any user part of a group, and accountability, ownership, access policies and rules can be assigned to a group.

- **Exploring metadata**: This activity in data governance comes under data definition. Most data governance tools support metadata management. Here, as shown in the following screenshot, the DataHub dashboard provides a summary of the metadata:

MySQL
347
AWS S3
7
Nifi
3
Domains
Sales
3
DataServices
2
Recently Viewed

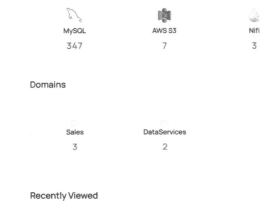

Figure 8.19 – Platform-wise and domain-wise resource summary

The preceding screenshot shows the various platforms and objects present on each platform. It also shows how many resources are in each domain. For enterprise data governance, it is important to monitor and understand how many resources are being used by different business units for auditing, tracking, and financial purposes. The following screenshot shows a MySQL table resource and its corresponding data definitions:

Figure 8.20 – Adding descriptions to metadata

As we can see, we can add a description to the metadata. On the other hand, the following screenshot shows how the schema of the table is loaded and seen in DataHub:

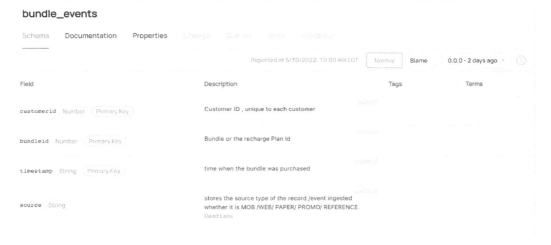

Figure 8.21 – Schema of the table

As we can see, descriptions, terms, and tags can be added and maintained in each column of the schema. This enables data definition activities for data governance.

- **Exploring lineage**: DataHub allows us to explore the lineage of the data. In the data definition of data governance, apart from maintaining the metadata and descriptions, proper governance must know the origin of the data and how it is being used. This aspect is covered by **Lineage**, as shown in the following screenshot:

Figure 8.22 – Data lineage

Here, we can see **1 Downstream** and no upstream events for the **chapter8input** S3 bucket. We can click on the **Visualize Lineage** button to see the lineage as a diagram, as shown in the following screenshot:

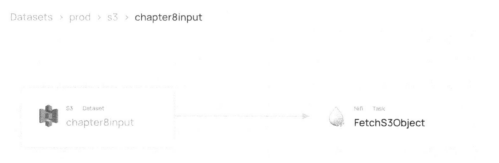

Figure 8.23 – Visualizing data lineage as a workflow

Apart from visual lineage, you can explore the impact analysis of this resource. This impact analysis helps communicate with the impacted groups in case a change or maintenance event occurs.

- **Creating accountability by adding a domain and owners to a resource**: We can start creating accountability of resources by attaching a domain and one or multiple owners to a resource from the landing page of the resource, as shown in the following screenshot:

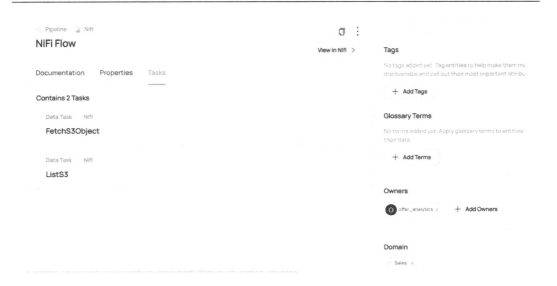

Figure 8.24 – Assigning a domain and owners

An owner can be a group or a user. However, it is advisable to assign ownership to a group and add users to that group. When adding an owner, you can choose the type of ownership, as shown in the following screenshot:

Figure 8.25 – Adding an owner

Here, while adding the **offer_analytics** owner, we set the type of owner to **Technical Owner**. An owner can be one of three types – **Technical Owner**, **Business Owner**, or **Data Steward**.

Technical Owner is accountable for producing, maintaining, and distributing the asset. **Business Owner** is a domain expert associated with the asset. Finally, **Data Steward** is accountable for the governance of the asset.

- **Setting policies**: DataHub allows us to set policies. Policies are sets of rules that define privileges over an asset or the DataHub platform. There are two categories of policies – **Platform** and **Metadata**. The **Platform** policy allows us to assign DataHub platform-level privileges to users or groups, while the **Metadata** policy allows us to assign metadata privileges to users or groups. To create a new policy, navigate to **Settings | Privileges**. Upon clicking the **Create New Policy** button, a wizard appears, as shown in the following screenshot:

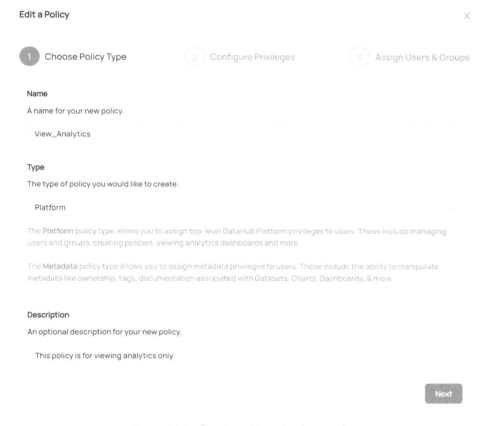

Figure 8.26 – Creating a View_Analytics policy

Here we are creating a new policy called **View_Analytics**. We have chosen **Platform** as the type of policy. Optionally, we can add a description of the policy. Here, we have added a description that states **This policy is for viewing analytics only**.

Click **Next** to go to the **Configure Privileges** section, as shown in the following screenshot:

Figure 8.27 – Configuring policy privileges

Here, we are configuring the platform privileges for the policy that we are defining. Finally, we must select/specify the users or groups that this policy will be applied to. The following screenshot shows how we can assign users or groups to a policy:

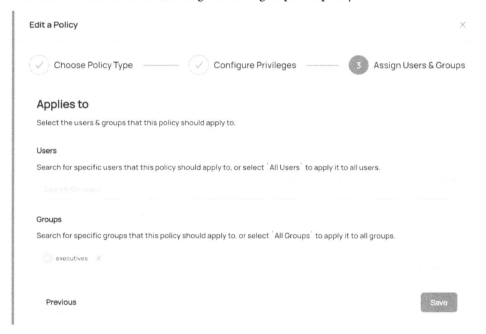

Figure 8.28 – Assigning users/groups to a policy

As we can see, the **View_Analytics** policy is assigned to the **executives** group.

- **Visual analytics**: DataHub also allows users to create and save dashboards for the analytics required for the data governance of an organization. The following screenshot shows various metrics, such as the most viewed datasets and data governance completeness (how much of the documentation/lineage/schema is well defined) for different entity types:

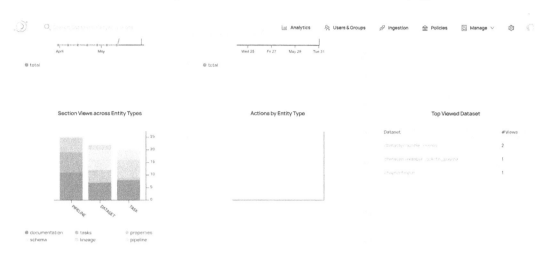

Figure 8.29 – Various visual analytics metrics

The following screenshot shows some other dashboard charts where we can find the platforms that were used per domain and their count. We can also see the number of entities (datasets or pipeline components) per platform:

Data Landscape Summary

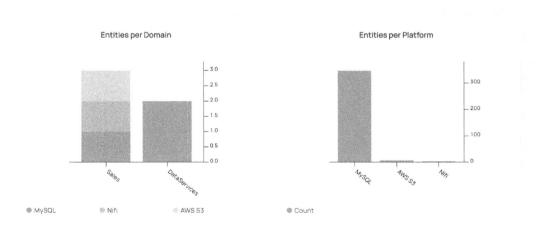

Figure 8.30 – Data Landscape Summary

Now that we have discussed and learned about the concepts of data governance and have practically implemented data governance around a use case, let's learn about data security.

Understanding the need for data security

Before we understand the need for data security, let's try to define what data security is. **Data security** is the process of protecting enterprise data and preventing any data loss from malicious or unauthorized access to data. Data security includes the following tasks:

- Protecting sensitive data from attacks.

- Protecting data and applications from any ransomware attacks.

- Protecting against any attacks that can delete, modify, or corrupt corporate data.

- Allowing access and control of data to the necessary user within the organization. Again, read-only, write, and delete access is provided to the data based on the role and its use.

Some industries may have stringent data security requirements. For example, a US health insurance company needs to ensure PHI data is extremely well protected according to HIPAA standards. Another example is that a financial firm such as Bank Of America has to ensure card and account data is extremely secure because that can cause direct monetary loss. But even if there is no stringent data security requirement in your organization, data security is essential to prevent data loss and loss of trust of the organization's customers.

Although data privacy and data security are overlapping terms, it helps to understand the subtle difference between the two. Data privacy ensures that only an authenticated user can access the data, and even if the data is accessed somehow, it should be encrypted or tokenized so that unauthorized users cannot use the data. Data security includes data privacy concerns, but apart from that, it focuses on tackling any malicious activity concerning the data. One simple example is that, to protect PHI data, Cigna Health insurance has encrypted all its sensitive data. This ensures data privacy but not data security. Although hackers may not be able to decipher the encrypted data, they can still delete the data or double encrypt the data so that the data becomes unusable.

Now, let's discuss why data security is needed.

Every year, the damage that's caused by data breaches is around $8 billion in the US alone and, on average, each incident causes approximately 25,000 accounts to be compromised. Let's explore a few of the biggest data breaches that have happened over the last few years:

- In June 2013, Capital One reported a data breach that affected all 106 million of its accounts. Personal information such as credit history, social security numbers, and bank accounts was compromised.

- In June 2021, a massive data breach happened on LinkedIn, wherein 700 million accounts (92% of the total accounts at that time) were affected. User data was posted for sale on the dark web.

- In April 2019, Facebook reported an attack that compromised 533 million accounts.

As we can see, these breaches not only pose security risks to the compromised data but also damage the reputation and trust of the company. Now, let's discuss a few common data security threats, as follows:

- **Phishing and other social engineering attacks**: Social engineering is a common way to trick or manipulate individuals to gain unauthorized access to corporate data or gather confidential information. Phishing is a common form of social engineering attack. Here, messages or emails that appear to be from a trusted source are sent to individuals. They contain malicious links that, when clicked, can give unauthorized access to corporate networks and data.

- **Insider threats**: Insider threats are caused by an employee who inadvertently or intentionally threatens the security of corporate data. They can be of the following types:

  - Malicious actors, who intentionally steal data or cause harm to the organization for personal gain.

  - Non-malicious actors, who cause threats accidentally or because they are unaware of security standards.

  - Compromised users, whose systems are compromised by external attackers without their knowledge. Then, attackers perform malicious activities, pretending to be legitimate users.

- **Ransomware**: This is a form of malware that infects corporate systems and encrypts data so that it becomes useless without a decryption key. Attackers then display a ransom message for payment to release the decryption key.

- **SQL injection**: Here, attackers gain access to databases and their data. Then, they inject unwanted code into seemingly innocent SQL queries to perform unwanted operations, thereby either deleting or corrupting the data.

- **Distributed Denial of Service (DDoS) attacks**: A DDoS attack is a malicious attempt to disrupt a web resource such as a web service. It acts by sending a huge volume of dummy calls to the web server. Due to the extreme volume of service calls at a very short burst, the web server crashes or becomes unresponsive, causing web resource downtime.

- **Compromised data in the cloud**: Ensuring security in the cloud is a big challenge. Security is required while sending data over the network and for data at rest in the cloud.

In this section, we discussed what data security is and the threats that can occur in absence of it by covering some real-world examples. Now, let's discuss the solutions and tools available for data security.

Solution and tools available for data security

In the previous section, we briefly discussed what data security is and why it is needed. We also looked at a few common data security threats. The solutions and tools described here help mitigate or minimize the risk from the threats discussed in the previous section:

- **Data discovery and classification**: To ensure data security, it is important to discover sensitive information. This technique uses data discovery to classify data into various security labels (such as confidential, public, and so on). Once classification is done, security policies can be applied to the various classifications of data according to the organization's needs.

- **Firewalls**: This is the first line of defense against any network intrusions. They exclude any undesirable traffic from entering the network. They also help open specific ports to the external network, which gives hackers less of a chance to enter the network.

- **Backup and recovery**: This helps organizations protect themselves in case the data is deleted or destroyed.

- **Antivirus**: This is used to detect any viruses, Trojans, and rootkits that can steal, modify, or damage your data. They are widely used for personal and corporate data security.

- **Intrusion Detection and Prevention Systems (IDS/IPS)**: IDS/IPS performs a deep inspection of packets on network traffic and logs any malicious activities. These tools help stop DDoS attacks.

- **Security Information and Event Management (SIEM)**: This analyzes recorded logs from various kinds of devices such as network devices, servers, and applications, and generates security alert events based on specific criteria and thresholds.

- **Data Loss Prevention (DLP)**: This mechanism monitors different devices to make sure sensitive data is not copied, moved, or deleted without proper authorization. It also monitors and logs who is using the data.

- **Access control**: This technique allows or denies read, write, or delete access to individual data resources. Access controls can be implemented using an **Access Control List (ACL)** or **Role-Based Access Control (RBAC)**. Cloud security systems such are IAM are used to enforce access control on the cloud.

- **Security as a Service**: This is modeled on the lines of Software-as-a-Service. Here, service providers provide the security of corporate infrastructure on a subscription basis. It uses a pay-as-you-go model.

- **Data encryption**: Data such as PHI can be very sensitive. If it is lost or leaked, this can cause regulatory issues and heavy financial losses. To provide data protection on such data, we can either encrypt the data (where the masked data loses all the original data properties, such as size) or tokenize it (where the masked data retains the properties of the original data).

- **Physical security**: Finally, physical security is essential to stop unauthorized access to data. Physical security can be enabled by creating strong security policies and implementing them. Policies such as locking the system whenever you step out, no tailgating, and others can help prevent social engineering attacks and unauthorized access to data.

In this section, we learned about the data security solutions and tools that are available in the industry. Now, let's summarize what we've learned in this chapter.

Summary

In this chapter, we learned what data governance is and why is it needed. Then, we briefly discussed the data governance framework. After that, we discussed how to develop a practical data governance solution using DataHub. Next, we learned what data security is and why it is needed. Finally, we briefly discussed the various solutions and tools that are available for ensuring data security.

With that, we have learned how to ingest data using real-time and batch-based pipelines, popular architectural patterns for data ingestion, and data governance and data security. In the next chapter, we will discuss how to publish the data as a service for downstream systems.

Section 3 – Enabling Data as a Service

This section of the book focuses on architecting solutions for Data as a Service. In this part, you will learn how to build various kinds of Enterprise grade **Data as a Service (DaaS)** solutions and secure and manage them properly.

This section comprises the following chapters:

9

Exposing MongoDB Data as a Service

In the previous chapters, we learned how to analyze and design solutions for various data ingestion and storage problems. We also learned how to analyze and classify those problems. After that, we learned how to apply scalable design principles and optimally choose technologies to implement those solutions. Finally, we learned how to develop, deploy, execute, and verify those solutions. However, in a real-world scenario, it is not always a good idea to expose the whole database to downstream systems. If we plan to do so, we must ensure that proper authorization and access rules are implemented on the database (please refer to the *Publishing problems* section of *Chapter 1*, *Basics of Modern Data Architecture*, for various ways of publishing data). One of the ways to give selective and authorized access to data is by publishing via **Data as a Service (DaaS)**.

DaaS enables data to be published by a platform and language-independent web service such as SOAP, REST, or GraphQL. In this chapter, we will analyze and implement a DaaS solution using the REST API to publish the previously ingested and sorted data from a MongoDB database. Here, we will learn how to design, develop, and unit test a REST application for publishing data. We will also learn how to deploy our application in Docker. In addition to this, we will briefly learn how to use API management tools and how they can be of help. Although Apigee is the most popular API management tool, we will be using AWS API Gateway as we are deploying and running our application on an AWS cluster. By the end of this chapter, you will know what DaaS is and when it should be used. You will also know how to design and develop a DaaS API, as well as how to enable security and monitor/control traffic on a DaaS API using API management tools.

In this chapter, we're going to cover the following main topics:

- Introducing DaaS – what and why
- Creating a DaaS to expose data using Spring Boot
- API management
- Enabling API management over the DaaS API using AWS API Gateway

Technical requirements

To complete this chapter, you will need the following:

- Prior knowledge of Java
- OpenJDK 1.11 installed on your local system
- Maven, Docker, and Postman installed on your local system
- An AWS account
- The AWS CLI installed on your local system
- IntelliJ Idea Community or Ultimate Edition installed on your local system

The code for this chapter can be found in this book's GitHub repository: `https://github.com/PacktPublishing/Scalable-Data-Architecture-with-Java/tree/main/Chapter09`.

Introducing DaaS – what and why

In the introduction, we briefly discussed and established that DaaS is useful for publishing already-ingested and analyzed data securely.

But *what* is DaaS? It is a data management strategy that enables data as a business asset, which makes valuable and business-critical data accessible on demand to various internal and external systems. **Software-as-a-Service** (**SaaS**) started becoming popular in the late 90s when software was provided to consumers on demand. Similarly, DaaS enables access to data on demand. With the help of **service-oriented architectures** (**SOAs**) and APIs, it enables secure platform-independent data access. The following diagram provides an overview of a DaaS stack:

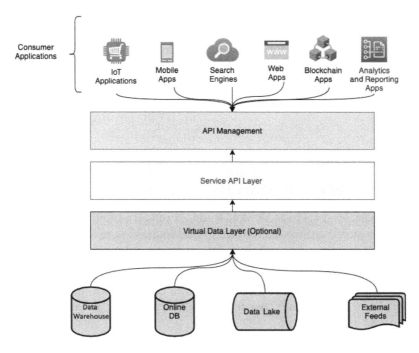

Figure 9.1 – DaaS stack overview

As we can see, data from different kinds of data stores, such as data warehouses, data lakes, or online databases, can be unified by a virtual data layer that can be used to build the services or API layer. The API management layer is present between the users of the data and the API layer. The API management layer is responsible for registering, securing, documenting, and orchestrating the underlying Service APIs. All consumer applications interact with the API management layer to consume the data from the Service APIs.

Now that we know what a DaaS is, let's figure out *why* it's important.

Data is the new gold. Every organization collects huge volumes of data, but a successful organization knows how to optimally use that data to drive its profits. Data collection and storage are done using different mediums across the organization. But to fully utilize the potential of that data, teams across the organization should be able to access it easily and securely. Data in silos, maintained by a team, may have the opportunity to derive value across the organization.

On the other hand, exposing a database to teams across the organization may be a governance or security headache. This is where DaaS plays a vital role. By exposing valuable and business-critical data using APIs, DaaS enables teams to share data with internal teams while implementing the necessary access and security checks on the dataset. This helps teams easily register and subscribe to data on demand instead of having to go through the unnecessary and cumbersome effort of analyzing, ingesting, and maintaining a dataset that is already ingested and maintained by some other team in the organization. This enables fast-paced development and cost saving in developing and maintaining redundant datasets across the organization. In addition to this, DaaS has enabled a lot of businesses to selectively provide data on demand to the external world for profit or usability.

According to Gartner's hype cycle, DaaS is still a long way from reaching its plateau of peak productivity, which means it has the potential to be one of the most impactful advancements in data engineering for the next decade.

Benefits of using DaaS

The following are a few of the major benefits of using DaaS:

- **Agility**: DaaS enables users to access data without needing a comprehensive understanding of where data is stored or how is it indexed. It also enables teams to focus on their business functionality, and not unnecessarily spend time storing and managing the data. This helps considerably decrease time-to-market.

- **Easy maintainability**: There are fewer maintenance headaches for the teams that use DaaS to get their data since they don't have to worry about managing and maintaining it, nor its storage and data pipelines.

- **Data quality**: Since data is served by APIs, data is more non-redundant, so taking care of data quality becomes much easier and more robust.

- **Flexibility**: DaaS gives companies the flexibility to trade off between initial investment versus operational expenses. It helps organizations save costs on the initial setup to store data, as well as ongoing maintenance costs. It also enables teams to get data on demand, which means that teams don't need to have a long-term commitment to that service. This enables teams to start using the data provided by DaaS in a much quicker fashion. On the other hand, if, after a while, there is no need for that data or the user wants to move to a newer technology stack, the migration becomes much faster and hassle-free. DaaS also enables easy integration in on-premises environments or the cloud for its users.

Now that we have understood the concept and benefits of publishing data using a DaaS, in the next section, we will learn how to implement a DaaS solution using the REST API.

Creating a DaaS to expose data using Spring Boot

In this section, we will learn how to expose a REST-based DaaS API using Java and Spring Boot. But before we try to create the solution, we must understand the problem that we are going to solve.

> **Important note**
> **REST** stands for **Representational State Transfer**. It is not a protocol or standard; instead, it provides certain architectural constraints to expose the data layer. The REST API allows you to transfer the representational state of a data resource to the REST endpoint. These representations can be in JSON, XML, HTML, XLT, or plain text format so that they can be transferred over the HTTP/(S) protocol.

Problem statement

In the solution described in *Chapter 6, Architecting a Real-Time Processing Pipeline*, we analyzed and ingested analytical data in a MongoDB-based collection. Now, we want to expose the documents present in the collection using a DaaS service that can be searched for by either `ApplicationId` or `CustomerId`. In the next section, we will analyze and design the solution and then implement it.

Analyzing and designing a solution

Let's analyze the requirement to solve the given problem. First, we will note down all the facts and information available. The following are the facts we know:

- The data to be published is stored in a MongoDB collection hosted in the cloud
- We need to publish the DaaS with security and API management
- We need to create endpoints so that the data can be fetched by either `ApplicationId` or `CustomerId`
- The DaaS that's been built for this data should be platform or language-independent

Based on these facts, we can conclude that we don't necessarily need a virtual data layer. However, we need the application to publish two endpoints – one exposing data based on `ApplicationId` and the other exposing data based on `CustomerId`. Also, since the MongoDB instance is in the cloud, and most of the frontend applications using this DaaS are in the cloud, it makes sense to deploy this application in an EC2 or **Elastic Container Repository** (**ECR**) instance. However, since we have no information about the traffic that will use this DaaS, it makes more sense to containerize the application so that, in the future, we can easily scale out the application by adding more containers.

The following diagram depicts the proposed solution architecture for our problem:

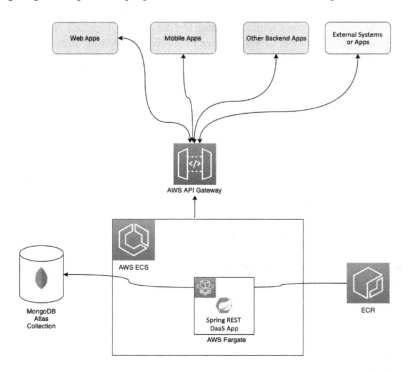

Figure 9.2 – Proposed solution architecture of DaaS

As we can see, we use our REST application to read data from MongoDB and fetch the results for the REST endpoint. We use Spring Boot as the technology stack to build the REST API as it is modular, flexible, extendible, and provides an amazing range of I/O integrations and community support. We will create Docker containers for this application and deploy them in the AWS Elastic Container Service cluster using the AWS Fargate service. This gives us the flexibility to scale up or down quickly when traffic increases in the future. Finally, we apply the API management layer on top of the deployed application. There are many API management tools available on the market, including Google's APIJEE, Microsoft's Azure API Management, AWS API Gateway, and IBM's API Connect. However, since our stack is deployed on AWS, we will be using the native API management tool of AWS: AWS API Gateway.

Now that we have discussed the overall architecture of the solution, let's learn more about the design of the Spring Boot application. The following diagram depicts the low-level design of the Spring Boot application:

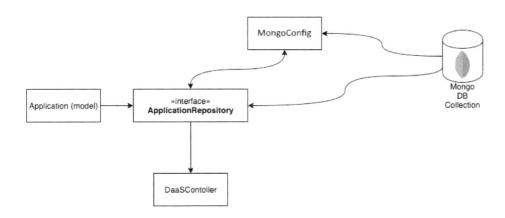

Figure 9.3 – Low-level design of the Spring Boot REST application

Our Spring application has a controller class called `DaaSController` that exposes two GET endpoints, as follows:

- GET /rdaas/application/{applicationId}
- GET /rdaas/customer/{id}/application

The first REST endpoint returns the application document based on `applicationId`, while the second REST endpoint returns all the applications of a given customer using `customerId` as the search criteria.

The `MongoConfig` class is used to configure and initialize `mongoTemplate` and `MongoRepository`. We create a custom `MongoRepository` called `ApplicationRepository` that uses `QueryDSL` to dynamically generate custom MongoDB queries at runtime. The controller class uses the `ApplicationRepository` class to connect to MongoDB and fetch document(s) from the collection as per the request.

With that, we have analyzed the problem and created a solution design. Now, let's discuss how we can implement this solution. First, we will develop the Spring Boot REST application.

Implementing the Spring Boot REST application

In this section, we will learn how to build the Spring Boot application to implement the solution we designed in the preceding section.

First, we must create a Maven project using our IDE and add the following Spring dependencies:

```
<dependency>
    <groupId>org.springframework.boot</groupId>
```

```xml
    <artifactId>spring-boot-starter-web</artifactId>
</dependency>
<dependency>
    <groupId>org.springframework.boot</groupId>
    <artifactId>spring-boot-starter-test</artifactId>
    <scope>test</scope>
</dependency>
<dependency>
    <groupId>org.springframework.boot</groupId>
    <artifactId>spring-boot-autoconfigure</artifactId>
</dependency>
```

These dependencies ensure that all the Spring Boot basic dependencies, as well as the REST-based dependencies, are fulfilled. However, we must add dependencies related to MongoDB. The following dependencies are related to Spring's MongoDB integration, as well as the necessary QueryDSL dependencies, to write custom MongoDB queries via the Spring JPA implementation:

```xml
<!-- mongoDB dependencies -->
<dependency>
    <groupId>org.springframework.boot</groupId>
    <artifactId>spring-boot-starter-data-mongodb</artifactId>
</dependency>


<!-- Add support for Mongo Query DSL -->

<dependency>
    <groupId>com.querydsl</groupId>
    <artifactId>querydsl-mongodb</artifactId>
    <version>5.0.0</version>
    <exclusions>
        <exclusion>
            <groupId>org.mongodb</groupId>
            <artifactId>mongo-java-driver</artifactId>
        </exclusion>
    </exclusions>
</dependency>
<dependency>
```

```
    <groupId>com.querydsl</groupId>
    <artifactId>querydsl-apt</artifactId>
    <version>5.0.0</version>
</dependency>
<!-- https://mvnrepository.com/artifact/javax.annotation/javax.
annotation-api -->
<dependency>
    <groupId>javax.annotation</groupId>
    <artifactId>javax.annotation-api</artifactId>
    <version>1.3.2</version>
</dependency>
```

Apart from these dependencies, we need to add build plugins to the pom.xml file. These plugins help generate Q classes dynamically, which are required for QueryDSL to work properly. The following plugins need to be added:

```
<plugins>
    <plugin>
        <groupId>org.springframework.boot</groupId>
        <artifactId>spring-boot-maven-plugin</artifactId>
    </plugin>

    <!-- Add plugin for Mongo Query DSL -->

    <plugin>
        <groupId>com.mysema.maven</groupId>
        <artifactId>apt-maven-plugin</artifactId>
        <version>1.1.3</version>
        <dependencies>
            <dependency>
                <groupId>com.querydsl</groupId>
                <artifactId>querydsl-apt</artifactId>
                <version>5.0.0</version>
            </dependency>
        </dependencies>
        <executions>
            <execution>
                <phase>generate-sources</phase>
```

```
                <goals>
                    <goal>process</goal>
                </goals>
                <configuration>
                    <outputDirectory>target/generated-sources/
apt</outputDirectory>
                    <processor>org.springframework.data.
mongodb.repository.support.MongoAnnotationProcessor</processor>
                    <logOnlyOnError>false</logOnlyOnError>
                </configuration>
            </execution>
        </executions>
    </plugin>
</plugins>
```

Now that we have added all the necessary dependencies, we will create the entry point, or the `main` class, of our Spring Boot application, as follows:

```
@SpringBootApplication(scanBasePackages = "com.
scalabledataarch.rest")
public class RestDaaSApp {

    public static void main(String[] args) {
        SpringApplication.run(RestDaaSApp.class);
    }
}
```

As per the preceding code, all the Bean components in the `com.scalabledataarch.rest` package will be scanned recursively and instantiated when the Spring Boot application is started. Now, let's create Mongo configuration beans using the `Configuration` class called `MongoConfig`. The source code for the same is as follows:

```
@Configuration
@EnableMongoRepositories(basePackages = "com.scalabledataarch.
rest.repository")
public class MongoConfig {

    @Value(value = "${restdaas.mongoUrl}")
```

```
    private String mongoUrl;

    @Value(value = "${restdaas.mongoDb}")
    private String mongoDb;

    @Bean
    public MongoClient mongo() throws Exception {
        final ConnectionString connectionString = new
ConnectionString(mongoUrl);
        final MongoClientSettings
mongoClientSettings = MongoClientSettings.builder().
applyConnectionString(connectionString).serverApi(ServerApi.
builder()
                .version(ServerApiVersion.V1)
                .build()).build();
        return MongoClients.create(mongoClientSettings);
    }

    @Bean
    public MongoTemplate mongoTemplate() throws Exception {
        return new MongoTemplate(mongo(), mongoDb);
    }
}
```

As we can see, the MongoConfig class is annotated with @EnableMongoRepositories, where the base package of the repositories is configured. All classes extending the MongoRepository interface under the base package will be scanned and Spring beans will be created. Apart from that, we have created the MongoClient and MongoTemplate beans. Here, we used the com.mongodb. client.MongoClients API to create the MongoClient bean.

Next, we will create a model class that can hold the deserialized data from a MongoDB document. We can create the model Application class as follows:

```
import com.querydsl.core.annotations.QueryEntity;
import org.springframework.data.annotation.Id;
import org.springframework.data.mongodb.core.mapping.Document;

@QueryEntity
```

```
@Document(collection = "newloanrequest")
public class Application  {
    @Id
    private String _id;

    private String applicationId;
```

It is important to annotate the class with @Document and give the value of the collection name as its argument for spring-data-mongo to understand that this POJO represents a MongoDB document structure for the specified collection. Also, annotating with @QueryEntity is essential for QueryDSL to generate Q classes dynamically using apt-maven-plugin.

Now, we will use this Application POJO to write our custom Mongo repository, as shown in the following code:

```
public interface ApplicationRepository
extends MongoRepository<Application, String>,
QuerydslPredicateExecutor<Application> {

    @Query("{ 'applicationId' : ?0 }")
    Application findApplicationsById(String applicationId);

    @Query("{ 'id' : ?0 }")
    List<Application> findApplicationsByCustomerId(String id);


}
```

To implement a custom repository, it must implement or extend the MongoRepository interface. Since our ApplicationRepository uses QueryDSL, it must extend QuerydslPredicateExecutor. We can specify a Mongo query using the @Query annotation, as shown in the preceding code, which will be executed when the corresponding method is called.

Now, we will create a controller class called DaasController. The DaasController class should be annotated with the @RestController annotation to indicate that it is a Spring component that publishes REST endpoints. A basepath to the endpoints in DaaSController can be created using the @RequestMapping annotation, as shown in the following code snippet:

```
@RestController
@RequestMapping(path = "/rdaas")
public class DaasController {
...
}
```

Now, we will add our methods, each corresponding to a REST endpoint. The following code shows the source code for one of the methods:

```
...
@Autowired
ApplicationRepository applicationRepository;

@GetMapping(path= "/application/{applicationId}", produces =
"application/json")
public ResponseEntity<Application> getApplicationById(@
PathVariable String applicationId){
  Application application = applicationRepository.
findById(applicationId).orElseGet(null);
  return ResponseEntity.ok(application);
}
...
```

As we can see, the method that will be triggered when a REST endpoint call is made is annotated with @GetMapping or @PostMapping, based on the HTTP method type. In our case, we need a GET request. Each mapping should be accompanied by the URL path and the other necessary properties as parameters for these annotations. In this method, the autowired applicationRepository bean is used to fetch Mongo documents using the applicationId field.

Finally, we will create application.yml to set up the configuration parameters to run the Spring Boot application. The application.yml file in our case will look as follows:

```
restdaas:
  mongoUrl: <mongo url>
  mongoDb: newloanrequest
```

As shown in the source code of the application.yml file earlier, we configure various Mongo connectivity details in the application.yml file.

Now, we can run the application from our local machine by running the `main` class and test it using Postman, as follows:

1. First, click the + button in the **Collections** tab to create a new collection named `ApplicationDaaS`:

Figure 9.4 – Creating a new Postman collection

2. Add a request to the collection using the **Add request** option, as shown in the following screenshot:

Figure 9.5 – Adding a request to the Postman collection

3. Next, fill in the configurations for the HTTP method, REST URL, and headers. Now, you can execute the request using the **Send** button, as shown in the following screenshot:

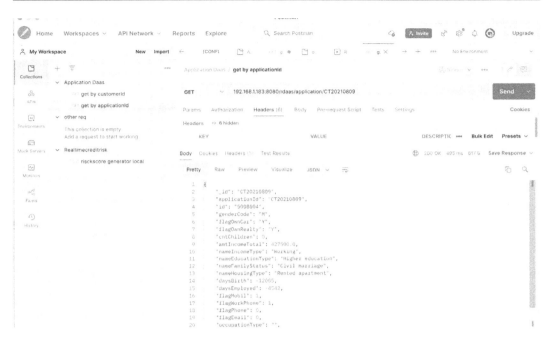

Figure 9.6 – Testing the REST API via Postman

Now that we have tested the application locally, let's see how we can deploy it in ECR. Follow these steps to deploy in ECR:

1. First, we need to containerize this application. To do that, we must create our Docker file. The source code for this `DockerFile` is as follows:

```
FROM openjdk:11.0-jdk
VOLUME /tmp

RUN useradd -d /home/appuser -m -s /bin/bash appuser
USER appuser

ARG JAR_FILE
COPY ${JAR_FILE} app.jar

EXPOSE 8080
ENTRYPOINT ["java","-jar","/app.jar"]
```

Here, the first line imports the base image, which contains the preconfigured OpenJDK 11 software. There, we create a volume called `/tmp`. Then, we add a new user to this Docker container called `appuser` using the `RUN useradd` command. Using the `USER` command, we log in as `appuser`. Then, we copy the JAR file to `app.jar`. The JAR file path is passed as an argument to this `DockerFile`. Passing the JAR file path as an argument will help us in the future if we want to build a **continuous integration and continuous deployment (CI/CD)** pipeline for this application. Then, port `8080` is exposed from the container. Finally, we run the `java -jar` command using the `ENTRYPOINT` command.

2. Now, we can build the Docker image by running the following command from a command line or Terminal:

    ```
    docker build -t apprestdaas:v1.0 .
    ```

3. This will create a Docker image called `apprestdaas` with a tag of `v1.0` in the local Docker image repository. You can either view the image listed in Docker Desktop or a Terminal by using the following command:

    ```
    docker images
    ```

Now that we have created the Docker image, let's discuss how can we deploy this image in an ECS cluster.

Deploying the application in an ECS cluster

In this section, we will discuss how to deploy our REST application in an AWS ECS cluster. Follow these steps:

1. First, we are going to create a repository in AWS ECR to store our Docker image. We will require the **Amazon Resource Name (ARN)** for this repository to tag and upload the image. First, we will navigate to ECR in the AWS Management Console and click **Create repository**. You'll see a **Create repository** page, as shown in the following screenshot:

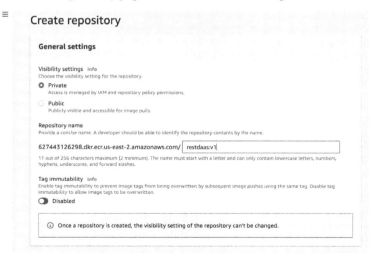

Figure 9.7 – The Create repository page

Here, fill in the repository's name, leave the rest of the fields as-is, and submit the request. Once it has been created, you will be able to see the repository listed on ECR's **Private repositories** page, as shown here:

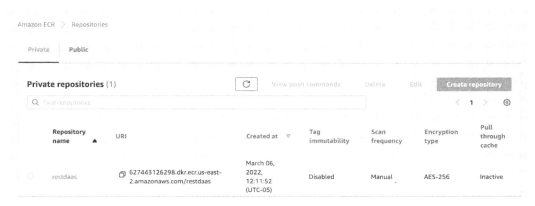

Figure 9.8 – The ECR repository has been created

2. Download the latest AWS CLI and install it. Then, configure the AWS CLI using the following command:

```
aws configure
```

While configuring the AWS CLI, you need to provide the access key ID and the secret access key. You can generate these variables by following the instructions at https://docs.aws. amazon.com/IAM/latest/UserGuide/id_credentials_access-keys. html#Using_CreateAccessKey.

3. Next, we need to generate an ECR login token for Docker using the following command:

```
aws ecr get-login-password --region <region>
```

When you run this command, an authentication token will be generated that is needed by Docker to push the image to ECR. We can pipe the previous command with the following command, where we directly pass the token to a docker login command. The final command looks like this:

```
aws ecr get-login-password --region <region> | docker
login --username AWS --password-stdin <accountid>.dkr.
ecr.<region>.amazonaws.com
```

In the preceding command, please replace region with the correct AWS region, such as us-east-2 or us-east-1, and accountid with the correct AWS account ID.

4. Next, we will tag the local Docker image with the ECR repository URI. This is required for ECR to map the correct repository with the image we are pushing. The command for this is as follows:

```
docker tag apprestdaas:v10 <accountid>.dkr.ecr.<region>.
amazonaws.com/restdaas:v1
```

5. Now, we will use the following command to push the Docker image into the ECR repository:

```
docker push <accountid>.dkr.ecr.us-east-2.amazonaws.com/
restdaas:v1
```

Upon running this command, the local Docker image will be pushed to the ECR repository with the `restdaas:v1` tag.

6. Now, let's create an AWS Fargate cluster. To do that, we must log into the AWS Management Console again. Here, search for `Elastic Container Services` and select it. From the **Elastic Container Service** dashboard, navigate to **Cluster** in the left pane and select **Create Cluster**:

Figure 9.9 – Creating an ECS cluster

7. Next, set the cluster template to **Networking only** and click **Next step**:

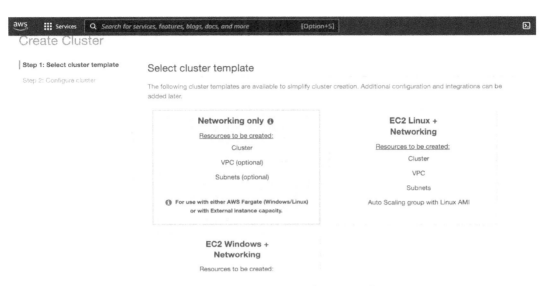

Figure 9.10 – Selecting an ECS cluster template

8. Then, enter the name of the cluster. Here, we will name it `daas-cluster`. Leave the other fields as-is. Now, click the **Create** button to create the new ECS cluster:

Figure 9.11 – Naming and creating the ECS cluster

9. Now, we will create an ECS task. From the dashboard of AWS ECS console, select **Task Definition** from the left menu and then click on **Create new Task Definition**, as shown in the following screenshot:

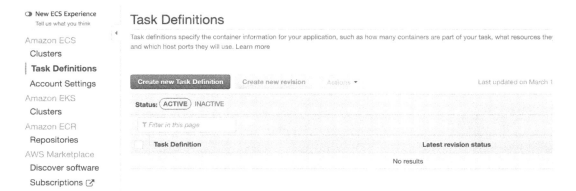

Figure 9.12 – Creating a new task definition

10. Then, under **Select launch type compatibility**, choose **FARGATE** and click **Next step**, as shown in the following screenshot:

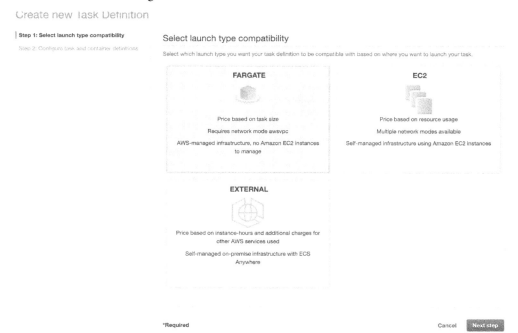

Figure 9.13 – Selecting the launch type for the ECS task

11. Next, we must set **Task definition name** to `restdaas`. Set **Task role** to **None** and **Operating system family** to **Linux**, as shown in the following screenshot:

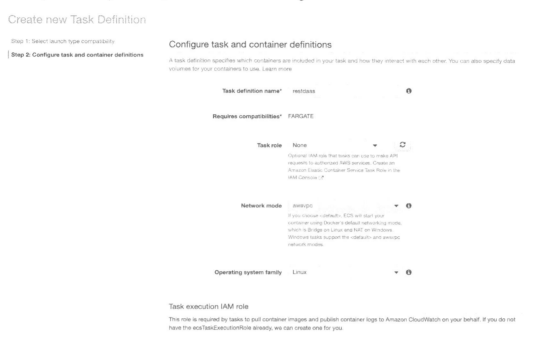

Figure 9.14 – Setting up a task definition

Leave **Task execution role** as-is, set **Task memory (GB)** to **1GB**, and set **Task CPU (vCPU)** to **0.5 vCPU**, as shown in the following screenshot:

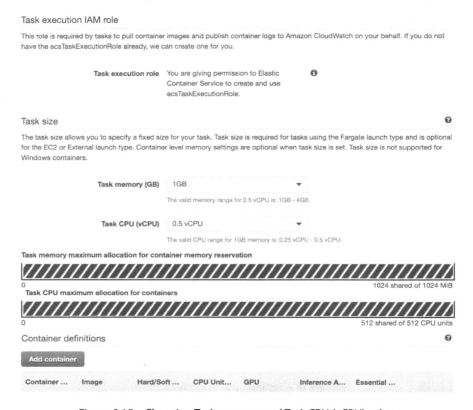

Figure 9.15 – Choosing Task memory and Task CPU (vCPU) values

12. Now, we will add a container to the ECS task. We can add the container by clicking on the **Add container** button shown in the preceding screenshot. A window will appear where we must enter `restdaas` as the container name and populate the ECR ARN of our image, as shown in the following screenshot:

Figure 9.16 – Adding a container to the ECS task

13. Click the **Add** button to add a container. Then, click the **Create** button on the **Create new Task Definition** page. This will create the new task, as shown in the following screenshot:

Figure 9.17 – ECS task created

As shown in the previous screenshot, the task we've created, **restdaas**, is in an **ACTIVE** state but it is not running.

14. Now, let's run the task. Click on the **ACTION** dropdown button and select **Run Task**. This submits the task so that it can be in runnable status. On clicking **Run Task**, a screen will appear where we must fill out various configurations for running the task, as shown in the following screenshot:

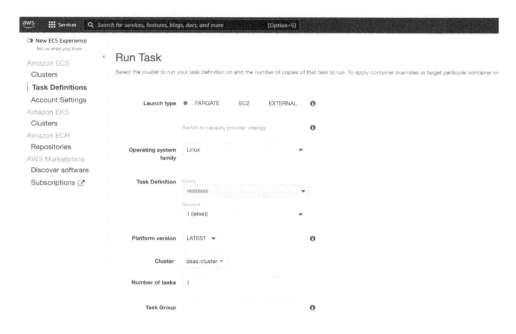

Figure 9.18 – Run Task

Set **Launch type** to **FARGATE** and **Operating system family** to **Linux**. Also, select any available VPC and subnet group, as shown in the following screenshot:

Figure 9.19 – Setting the VPC and security groups

15. As shown in the preceding screenshot, please make sure that **Auto-assign public IP** is set to **ENABLED**. Also, modify the security group so that you can add a custom TCP that allows port 8080 since our Spring Boot application will be running on port 8080, as shown in the following screenshot:

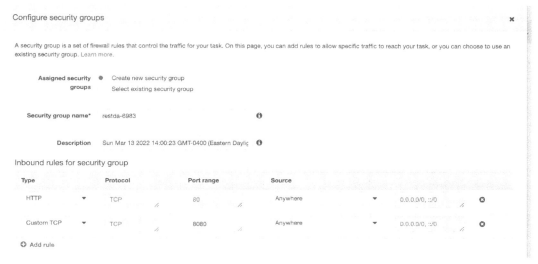

Figure 9.20 – Allowing port 8080

16. Leave the other fields as-is and click **Run Task**. By refreshing the **Task** tab, we can see that the task changes state from **PROVISIONING** to **RUNNING**. The following screenshot shows a **RUNNING** task:

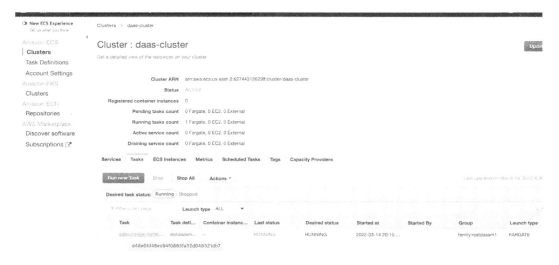

Figure 9.21 – Task in the RUNNING state

17. Now, we will test the DaaS that we've deployed in ECS. To do that, click on the text under the **Task** column shown in the preceding screenshot. This takes us to the **Running task** screen, as shown in the following screenshot:

Figure 9.22 – Running task details

As highlighted in the preceding screenshot, we will get the public IP. We will use this IP for testing purposes. Now, from Postman, we can test our REST endpoints using this IP address and port 8080, as shown in the following screenshot:

Figure 9.23 – Testing the REST endpoint that was deployed in AWS ECS

In this section, we learned how to develop a REST application to publish the DaaS, containerize the application, deploy it in an AWS ECS cluster, and test the endpoints. In the next section, we will understand the need for API management, and we will provide a step-by-step guide to attaching the API management layer to the REST endpoints that we developed and deployed in this section.

API management

API management is a set of tools and technologies that allows us to distribute, analyze, and control APIs that expose data and services across the organization. It can act like a wrapper on top of the APIs, whether they are deployed on-premises or in the cloud. It is always a good idea to use API management while we are architecting a solution to publish data via an API. First, let's understand what API management is and how it helps. The following diagram shows where and how the API management layer helps:

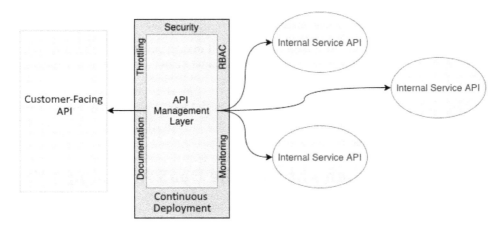

Figure 9.24 – How API management helps

As we can see, API management is a wrapper layer that sits between the customer-facing API and the internal service API. We can define resources and methods in the API management layer that get exposed to the customer. Primarily, an architecture gets the following benefits while using an API management layer:

- **Flexibility**: API management enables easy deployment in a different environment by enabling continuous deployment and testing. It also provides a unified interface where a single customer-facing API can be fetched from multiple complex internal service APIs. This enables easy integration and allows you to publish resources without the need to create extra APIs to integrate and manage multiple APIs. On the other hand, in a very dynamic technological landscape, chances are that internal service APIs may be updated or their structure may change frequently.

The API management layer gives us easy access to move from an older internal service API to a new one without changing or affecting the customer-facing API. This gives us huge flexibility in design and helps us overcome technical debts in the internal service API layer without any hassle.

- **Security**: API management provides security in different ways. First, it enables APIs to have custom authorizers such as OAuth. Second, it enables customer-specific usage plans and API keys. This ensures only a consumer who is registered with a usage plan and API key will be able to access the application. Also, it puts a limit on how many transactions a consumer can do per second. This, along with the throttling feature, helps us avoid any **distributed denial-of-service (DDoS)** attacks on the Service APIs. Apart from these, role-based access can be enabled using the RBAC feature of the API. All these security features make API management a necessary component in designing a DaaS architecture.

- **Documentation**: This allows you to easily create, publish, and maintain the documentation of the API. The published documentation can be easily accessed by consumers of the API, making their lives easy. Apart from this, even the *Swagger* and *OpenAPI* specifications can be published and maintained using API management.

- **Analysis**: One of the major advantages of using API management is the ability to monitor and analyze the traffic, latency, and other parameters while the API is deployed and used in production.

In this section, we understood what API management is and how it can help us create a robust architecture for DaaS solutions. In the next section, we will attach an API management layer on top of the ECS REST Service API that we developed earlier.

Enabling API management over the DaaS API using AWS API Gateway

In this section, we will discuss how to set up API management using AWS API Gateway. We will use the REST DaaS API that we developed and deployed in ECS earlier in this chapter. Follow these steps to set up an API management layer for our REST DaaS API:

1. In the AWS Management Console, search for AWS API Gateway and navigate to the **AWS API Gateway** service dashboard. From here, select **REST API** and click **Build**, as shown in the following screenshot:

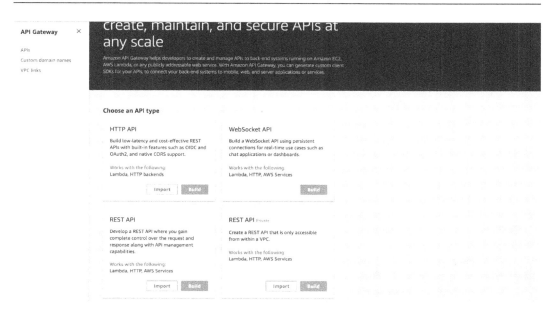

Figure 9.25 – The AWS API Gateway dashboard

2. A new window will open, as shown in the following screenshot. Select **REST** as the protocol and then select **New API** under **Create new API**. Fill in the API's name and a description and click **Create API**:

Figure 9.26 – Creating a REST API

3. Once the resource has been created, we will be taken to the details of the API. We can add resources or methods from the **Actions** dropdown in this interface, as shown in the following screenshot:

Figure 9.27 – Adding resources to the API

Here, we will click **Add Resource** and add a resource whose name and path are both `/loanapplications`. Then, we will add another resource under `loanapplications` whose name is `appId` and the path is `/{appId}`. Note that `{}` denotes that `appId` is a path variable. Finally, in the `appId` resource, we will add a method by selecting **Create Method**, as shown in the following screenshot:

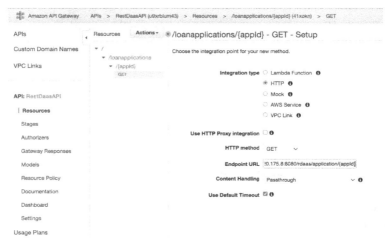

Figure 9.28 – Configuring the GET method

Set **Integration type** to **HTTP** and provide an **Endpoint URL**. Since our resource path has a path variable, the same path variable should be present in the **Endpoint URL** area as well. Here, we are mapping the /loanapplications/{appId} API resource to the /rdaas/ application/{appId} DaaS API resource.

4. Deploy the API by choosing the **Deploy API** option from the **Actions** dropdown, as shown in the following screenshot:

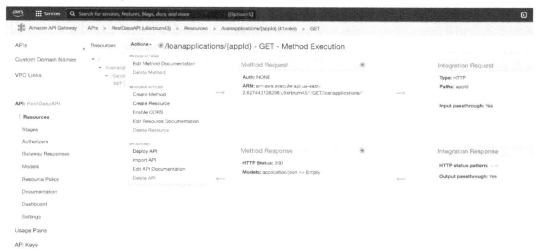

Figure 9.29 – Deploy API

Upon doing this, a window will appear, as shown in the following screenshot. Set **Deployment stage** to [**New Stage**] and fill in the stage's name and description. Finally, click **Deploy**:

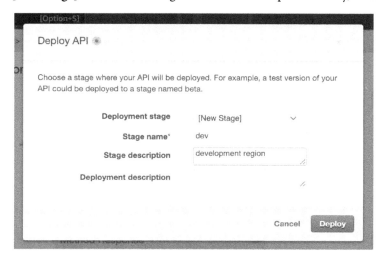

Figure 9.30 – Deploying the API configuration

5.   Now, we will test the new customer-facing API via Postman. But before that, we must find out the base URL of the API. To do so, navigate to **Stages** in the left pane and select **dev**. The **dev Stage Editor** page will appear, as shown in the following screenshot:

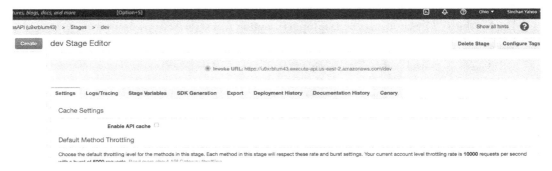

Figure 9.31 – Using dev Stage Editor to get the base URL of the API

As highlighted in the preceding screenshot, we can get the base URL.

Now, we can form the customer API by adding the consumer API's URI to the basepath; for example, `http://<baseurl>/loanapplications/<some_app_id>`. We can use this API and test it in Postman, as shown here:

Figure 9.32 – Testing the external API exposed by AWS API Gateway (without security)

We can also see the dashboard to monitor the API, as shown in the following screenshot:

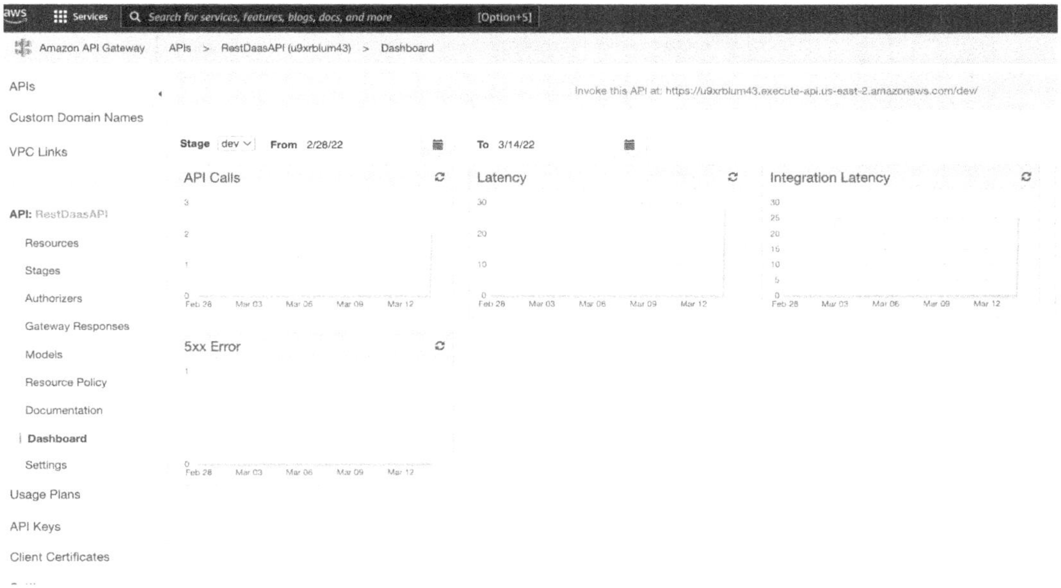

Figure 9.33 – AWS API Gateway dashboard for RESTDaasAPI

From this dashboard, we can gather useful information about the number of API calls made every day. We can also monitor the latency of the response or any internal server errors that have been noticed over time.

Now that we have added the API management layer, we will try to add API key-based security to the consumer-facing API.

6. Navigate to the **Resource** pane of **RestDaasAPI** and select the **GET** method under the {**appId**} resource. In the configuration, change the value of the required API key from **false** to **true**:

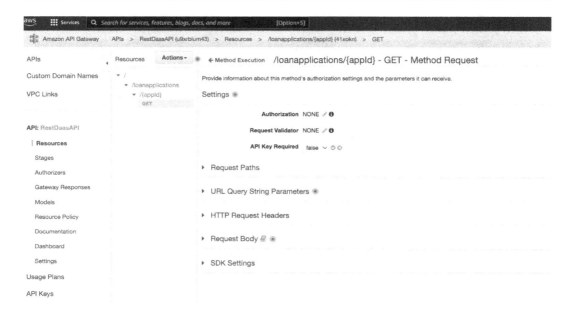

Figure 9.34 – Changing the API key's required value to true

Then, navigate to **Usage Plans** and create a usage plan. Set the API-level throttling parameters and monthly quota, as shown in the following screenshot:

Figure 9.35 – Create Usage Plan

Click **Next** and set the method-level throttling parameters, as shown here:

Figure 9.36 – Configuring the method-level throttling parameters

7. Click **Next** and set up a new API key for this usage plan by clicking the **Create API Key and add to Usage Plan** button:

Figure 9.37 – Generating a new API key to attach to the usage plan

8. Once you have clicked this, the **API Key** window will appear. Provide a name and description. Also, set **API key*** to **Auto Generate**:

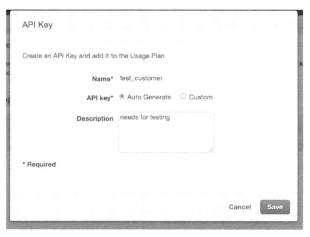

Figure 9.38 – The API Key window

9. Once we have saved the newly generated API key and created the usage plan, we can get the API key's value by navigating to **API Keys** and clicking on the **Show** option. By doing so, we can view the generated API key:

Figure 9.39 – Showing the generated API key by clicking Show (highlighted)

10. Once you have set up the API key as required in the API, and if you do not provide `apikey` in the header while invoking the REST endpoint, you will get an error message in the response body, similar to `{"message":"Forbidden"}`. Now, you must add a header called `x-api-key` whose value should be the API key that you generated in *Step 9*. Then, you can test the secure API via Postman, along with the API key, as shown in the following screenshot:

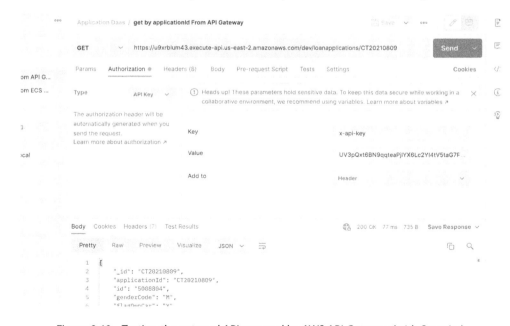

Figure 9.40 – Testing the external API exposed by AWS API Gateway (with Security)

In this section, we learned how to create an API management layer on top of our REST DaaS API. We also discussed how AWS API Gateway can help to monitor and secure the API.

Now, let's summarize what we've learned in this chapter.

Summary

In this chapter, we learned about the basics of DaaS. First, we discussed how to develop and test REST-based DaaS APIs using Spring Boot. Then, we learned how to containerize the application and publish the containers to the AWS ECR repository. We also learned how to deploy the containers published in the AWS ECR repository to an AWS ECS cluster. After that, we learned how to run this application using the cloud-managed Fargate service. Then, we learned about API management and its benefits. Finally, we implemented an API management layer to provide security and monitoring on top of our REST DaaS API using AWS API Gateway.

Now that we have learned how to build, deploy, publish, and manage a REST-based DaaS API, in the next chapter, we will learn how and when a GraphQL-based DaaS can be a good design choice. We will also learn how to design and develop a GraphQL DaaS API.

10

Federated and Scalable DaaS with GraphQL

In the previous chapter, we discussed how to publish ingested data in a format that is platform- and language-agnostic using the REST API. We also learned how to design and develop a **Data as a Service (DaaS)** layer using the REST API, as well as how to containerize and deploy the application on AWS ECS. Then, we learned what an API management system is and how it can help monitor and manage APIs more efficiently. Finally, we learned how to create an API management layer on top of our REST DaaS application using Amazon API Gateway.

In this chapter, we will learn how to implement DaaS using GraphQL instead of REST. To do so, we will learn what GraphQL is, and why and when it should be used. We will explore the benefits and shortcomings that GraphQL has concerning REST while discussing the various architectural patterns available for GraphQL-based solutions. Finally, we will learn about the power of federation in the GraphQL layer. By the end of this chapter, you should know about the basic concepts surrounding GraphQL and when to use this tool in data engineering solutions. You will also know how to design, implement, and test a GraphQL solution.

In this chapter, we're going to cover the following main topics:

- Introducing GraphQL – what, when, and why

- Core architectural patterns of GraphQL

- A practical use case – exposing federated data models using GraphQL

Technical requirements

For this chapter, you will need the following:

- Prior knowledge of Java

- OpenJDK-1.11 installed on your local system

- Maven installed on your local system

- GraphQL Playground installed on your local system

- IntelliJ Idea community or ultimate edition installed on your local system

The code for this chapter can be downloaded from this book's GitHub repository: `https://github.com/PacktPublishing/Scalable-Data-Architecture-with-Java/tree/main/Chapter10`.

Introducing GraphQL – what, when, and why

In this section, we will explore what GraphQL is. According to `graphql.org`, the official definition of GraphQL is that *"GraphQL is a query language for APIs and a runtime for fulfilling those queries with your existing data."* Let's dive a bit deeper to understand GraphQL.

Representational State Transfer (**REST**) has been the standard way of publishing data across systems, which is platform, device, and tool/language-agnostic. However, there are two major bottlenecks with REST:

- For fetching different related entities, we need multiple REST requests. We must also be mindful of different versions of the API. Having different endpoints and their versions for each entity of functionality is a maintenance headache.

- The request and response parameters are always fixed in REST. For example, there is a REST API that returns 100 fields. Suppose there is a consumer who only needs 10 fields. However, since responses are fixed, a REST request will always calculate and send all 100 fields. This, in turn, affects performance as it takes more time to form the response, as well as consumes more bandwidth to transfer a bigger payload.

GraphQL is the answer to overcome these challenges. GraphQL is an open standard or specification created by Facebook. It is a query language for APIs, where the client can query multiple entities and the desired fields from those entities while making the GraphQL request. The following diagram describes how GraphQL operates:

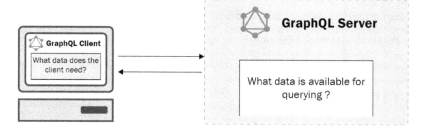

Figure 10.1 – How GraphQL operates

As shown in the preceding diagram, in GraphQL, it is the GraphQL client that defines what data it needs, while the GraphQL server publishes what data is available. So, essentially, GraphQL is a declarative way of fetching and updating data over APIs.

Let's try to understand this with an example. Since GraphQL was created by Facebook, we will take an example from a social networking use case. Let's suppose that we want to fetch users, their posts, and the comments related to that post. The following diagram shows how this can be designed using REST:

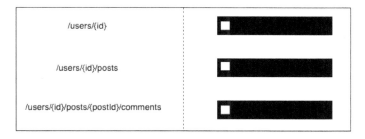

Figure 10.2 – A REST-based design requires three separate endpoints

As we can see, REST will have three endpoints – one for users, one for posts, and one for comments. To fetch a user's posts and comments on those posts, REST will require one call for users, one call for the posts, and one call per post for the comments. This is shown in the following diagram:

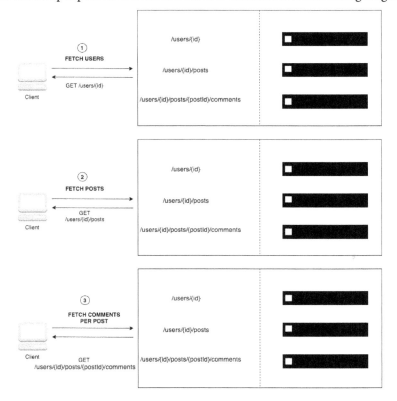

Figure 10.3 – Three separate calls are needed when using REST

As we can see, there are at least three REST calls (provided that the user has only one post) required for this use case to fetch data. Think of the number of calls required for a real user on a social platform. If the user has posted n number of posts, then the number of calls required to fetch this information will be $n+2$. That would seriously affect the UI experience and the performance of the website. However, in GraphQL, it only takes one call to fetch this information. The following diagram shows what a GraphQL request would look like:

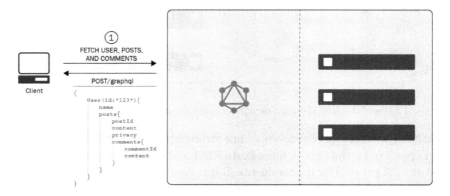

Figure 10.4 – A single GraphQL call can fetch all the required data

As you can see, the request payload of GraphQL declares the entities and fields that it needs to fetch. Therefore, the GraphQL client determines what data it needs.

Operation types

Now that we understand what GraphQL is, let's try to understand the various types of operations that GraphQL supports, as follows:

- **Queries**: These help query the API and only support data read operations. A sample query is as follows:

```
query myquery{
  byApplicationId(applicationId:"CT20210809"){
    applicationId
    id
    riskScore
  }
}
```

As shown in the preceding payload, you can optionally use a keyword query followed by the name that you want to assign to the query (here, it is myquery). byApplicationId is a query in GraphQL (like an endpoint in REST) that takes request arguments as parameters. Here, the byApplicationId query takes an argument called `applicationId`. Also, as you can see, the request contains the names of the fields that it wants to have returned, such as `applicationId`, id, and `riskscore`.

* **Mutations**: Mutations support both read and write operations. A sample query is as follows:

```
mutation updateApplicationMutation {
updateApplication(status:"closed") {
applicationId
custId
status
}
}
```

As shown in the preceding code, a mutation can be labeled with a mutation name using the `mutation` keyword. Here, it updates the application status in the database. So, it is used to write data.

* **Subscriptions**: In addition to queries and mutations, GraphQL also supports subscriptions. Like queries, they are used to fetch data, but they use long-lasting connections, which can change their result over time. This enables an event notification pattern, by which the server can push changes to the client. The following code shows what a subscription query looks like:

```
type subscription{
    commentsforPost(postId: "123ty4567"){
    commentId
    text
    ...
    }
}
```

Here, we are subscribing to the comments written for the post with a `postId`. Hence, a long-lasting connection is established between the client and the GraphQL server. The GraphQL server pushes any changes that are made on the comments to the client automatically.

Now, let's discuss the GraphQL schema. A GraphQL schema is a contract between the client and the server. It declares what operations and fields are available in GraphQL. It is strongly typed and is written using a standard **Schema Definition Language** (**SDL**). The following code block shows an example of a GraphQL schema:

```
type Application {
  applicationId: ID!
  id: String!
```

```
genderCode: String
cars: [String]
...
}
```

In the preceding code block, the `applicationId` field is of the `ID` data type, `genderCode` is of the `String` data type. and so on. In SDL, data types of GraphQL schemas are also known as scalars; the inbuilt data types are denoted as inbuilt scalars and customized data types are referred to as custom scalars. The exclamation mark after `ID` while defining `applicationId` denotes that it is a mandatory field. Similarly, `cars` is defined as a list since its data type is wrapped with `[]` (square brackets).

Now that we have internalized the basic concepts of GraphQL, we will explore why and when to use GraphQL.

Why use GraphQL?

In this section, we will explore the various benefits of GraphQL that make it a great solution. The advantages of GraphQL are as follows:

- **Strongly typed schema for a platform-independent API**: Usually, in the SOAP and REST APIs, our response is a platform-independent structure such as XML or JSON. However, neither of these formats is strongly typed. In GraphQL, each field in a schema must have a scalar type (it can either be an inbuilt scalar type or a customized one). This ensures GraphQL is less error-prone, more validated, and provides easy auto-completion capabilities for editors such as GraphQL Playground. The following screenshot shows the auto-completion suggestion provided by GraphQL Playground:

Figure 10.5 – Auto-completion feature in GraphQL Playground

- **No over-fetching or under-fetching**: In GraphQL, the client can only fetch the data that it requires. For example, if a GraphQL API supports hundreds of fields in its API response, the client doesn't need to fetch all those fields. If a client only needs 10 fields, the client can request the GraphQL API to send only those fields. However, if the same API is written in REST, even though the client requires only 10 fields, the response will return all the fields.

  Over-fetching is a common problem in REST, where, irrespective of the number of fields a client requires, it always fetches all the fields defined in the response body. For example, for a social networking site such as LinkedIn, a person's profile contains a huge number of columns, including demographic columns, skill set columns, awards and certification columns, and so on. A REST-based solution to solve this problem can be designed in two ways:

  - **Create a single API with all columns**: If we use this approach, a client requiring only demographic information will encounter an over-fetching problem.

  - **Create separate APIs for demographics, skill sets, awards and certifications, and so on**: Let's look at a scenario where the client needs all the information available. Multiple calls are required to fetch the data if we use this approach. This leads to under-fetching.

  So, we need a single solution that can solve these issues for both types of client requests. GraphQL solves this problem by allowing the client to choose which fields it wants to fetch.

- **Saves time and bandwidth**: GraphQL allows you to make multiple resource requests in a single GraphQL call, which saves a lot of time and bandwidth by reducing the number of network round trips to the GraphQL server. This is especially useful for improving the user experience and speed of fetching data on a client application.

- **No need for versioning**: In REST, when adding a new field or deleting an old field, this needs to be published as a new version to support consumer compatibility. With GraphQL, versioning is no longer required – one, because GraphQL supports partial data fetching from the response structure, and two, it supports publishing depreciation warnings to a depreciated field for the GraphQL client.

- **Schema stitching or combining multiple GraphQL schema**: GraphQL provides multiple ways to combine different GraphQL schemas and APIs into a single endpoint without much coding or implementation hassle. This feature helps develop a single, centralized GraphQL gateway. A GraphQL gateway enables multiple GraphQL APIs to be consumed from a single endpoint. It also enables the dynamic addition of newer GraphQL APIs in the future seamlessly. This makes GraphQL federated and scalable. Combining schemas can be achieved by technologies such as Apollo GraphQL Federation and Atlassian GraphQL Braids.

Now, let's see when we should use GraphQL.

When to use GraphQL

The following are a few scenarios where the GraphQL API is a better solution than the REST API:

- Apps where bandwidth usage matters, such as mobile apps or IoT device apps.

- Applications where there is a need for nested data to be fetched. GraphQL saves a lot of time and bandwidth, thus enabling better performance for GraphQL clients.

- Applications that publish DaaS. Here, this DaaS is consumed by multiple downstream teams with different data fetch requirements.

- When enabling GraphQL capabilities such as partial data fetching of the response and exposing a single endpoint by composing multiple APIs to improve the consumer experience for legacy REST-based applications.

In this section, we learned what GraphQL is and looked at typical use cases when GraphQL is used. In the next section, we will discuss the most popular GraphQL patterns that are used in the industry.

Core architectural patterns of GraphQL

In this section, we will discuss the various architectural patterns that are used for GraphQL. These patterns are independent of the technology used to implement it or the platform where it is deployed and executed. There are five different GraphQL patterns, as follows:

- **DaaS pattern**: Here, GraphQL server is used to expose the database layer. It can expose three operations – queries, mutations, and subscriptions (please refer to the *Operation types* section of this chapter). Using these operations, it can achieve **Create, Read, Update and Delete (CRUD)** operations like REST but also supports subscriptions on top of it. The following diagram shows this pattern:

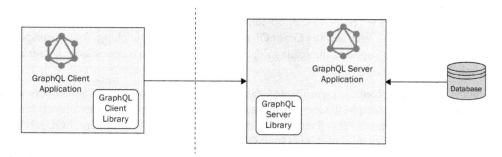

Figure 10.6 – DaaS pattern

As we can see, GraphQL exposes its queries and operations using the HTTP protocol. GraphQL provides server libraries in multiple languages, using which teams can build and run GraphQL Server applications. Also, GraphQL supports a variety of GraphQL client libraries in different languages. A list of supported languages is available at `https://graphql.org/code/`.

- **Integration layer pattern**: Here, GraphQL Server caters to data to provide access to multiple data sources in one go. This enables GraphQL to act like a data integration hub. The following diagram depicts how the integration layer pattern works:

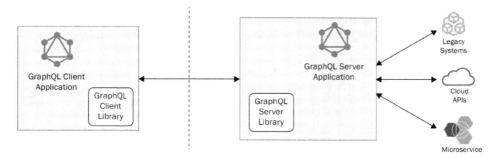

Figure 10.7 – Integration layer pattern

As we can see, GraphQL server is acting as an integration hub. It enables clients to make a single call, but the GraphQL server is fetching that data from different ecosystems such as microservices, legacy apps, and cloud APIs and sending a unified response to the client. This automatically reduces the complexity and number of calls that a GraphQL client must make.

- **Hybrid pattern**: The third GraphQL pattern is called the hybrid pattern because it explores a hybrid approach to the first two patterns. Here, GraphQL Server not only needs to be connected to microservices legacy systems, but also databases. The following diagram shows this pattern:

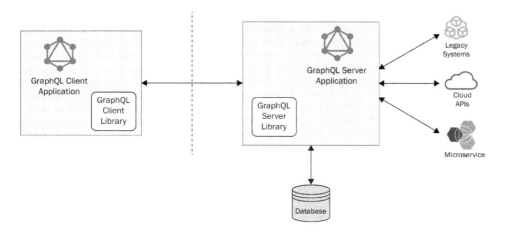

Figure 10.8 – Hybrid pattern

As we can see, GraphQL Server has its own database apart from the connections to different apps such as microservices and legacy systems. Therefore, GraphQL provides unified access to its client for different kinds of data sources when this pattern is used.

- **GraphQL with managed API**: To expose GraphQL APIs in an enterprise, it is necessary to have security and monitoring enabled. In this pattern, API Gateway provides monitoring, security, and throttling on GraphQL Server. The following diagram shows this pattern:

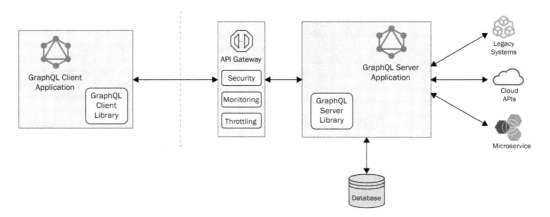

Figure 10.9 – GraphQL with managed API

- **Federated GraphQL pattern**: Here, a centralized GraphQL braid or federated GraphQL server is created. Other GraphQL nodes are connected to this GraphQL braid. Each of these nodes, in turn, fetches data from either a database, micro-service, or legacy app. The following diagram shows the federated GraphQL pattern:

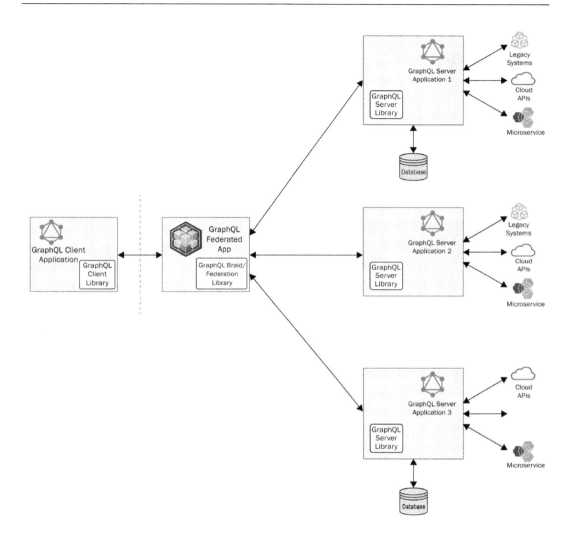

Figure 10.10 – Federated GraphQL pattern

The real power of this pattern is its amazing scalability and data federation. Newer nodes can be seamlessly added at any time to the GraphQL braid, without any application downtime.

In this section, we learned about the various core GraphQL patterns, how they operate, and when they are useful. In the next section, we will learn how to develop a GraphQL server application.

A practical use case – exposing federated data models using GraphQL

In this section, we will learn how to develop DaaS using GraphQL in Java. To implement the solution, we will publish the same set of APIs that we published earlier using REST, but this time, we will implement the solution using GraphQL.

Before we start implementing GraphQL, it is important to design the GraphQL schema for our use case. In our use case, we need to read credit card applications from MongoDB using either the application ID or consumer ID. This was why we needed two separate endpoints in the REST-based solution (please refer to *Chapter 9, Exposing MongoDB Data as a Service*, for the REST-based DaaS solution).

Let's analyze the requirements from a different perspective – that is, while considering the GraphQL-based solution. The biggest difference that GraphQL makes is that it reduces the number of endpoints, as well as the number of calls. So, for our use case, we will have a single endpoint. Also, according to our use case, we are only interested in fetching data. Hence, we will only be using the Query type of operation. To support multiple functionalities in GraphQL, we must have multiple such fields in the query that can take arguments such as the following:

```
type Query{
    customerById(id: String):Customer
    customerByName(firstname: String, lastName: String):
[Customers]
}
```

In our case, we need two such fields – byApplicationId and byCustomerId – both of which should return a custom type called Application. The following code snippet shows part of our GraphQL schema:

```
type Query{
    byApplicationId(applicationId: ID):Application
    byCustomerId(custId: String):[Application]
}
```

As shown in the preceding code block, byApplicationId always returns only one Application, as applicationId is a primary key. Therefore, byApplicationId is of the Application type. However, since there can be multiple applications for the same customer, byCustomerId is of the [Application] type, which denotes a list of Application. Now, let's define the type – Application – in the GraphQL schema. The following code block shows the SDL for the Application type:

```
type Application {
    applicationId: ID!
    id: String!
    genderCode: String
    flagOwnCar: String
    flagOwnRealty: String
    cntChildren: Int
    amtIncomeTotal: Float
    nameIncomeType: String
    nameEducationType: String
    nameFamilyStatus: String
    nameHousingType: String
...
}
```

Here, in the SDL for the Application type, applicationId is of the ID type, which denotes that it is the unique key for the Application type. Also, the exclamation mark (!) seen in the applicationId and id fields denotes that those fields are non-nullable. The complete schema is available at https://github.com/PacktPublishing/Scalable-Data-Architecture-with-Java/blob/main/Chapter10/sourcecode/GraphQLDaas/src/main/resources/schema.graphqls.

To create a Spring Boot Maven project and add the required Maven dependencies, the following Maven dependencies should be added in the pom.xml file:

```
<dependency>
    <groupId>org.springframework.boot</groupId>
    <artifactId>spring-boot-starter-web</artifactId>
</dependency>
<dependency>
    <groupId>org.springframework.boot</groupId>
    <artifactId>spring-boot-starter-test</artifactId>
    <scope>test</scope>
</dependency>
<dependency>
    <groupId>org.springframework.boot</groupId>
    <artifactId>spring-boot-autoconfigure</artifactId>
</dependency>
```

Apart from this, the following dependencies should be added to support MongoDB-related dependencies, as well as QueryDSL-related dependencies, in the Spring Boot application:

```xml
<!-- mongoDB dependencies -->
<dependency>
    <groupId>org.springframework.boot</groupId>
    <artifactId>spring-boot-starter-data-mongodb</artifactId>
</dependency>


<!-- Add support for Mongo Query DSL -->

<dependency>
    <groupId>com.querydsl</groupId>
    <artifactId>querydsl-mongodb</artifactId>
    <version>5.0.0</version>
    <exclusions>
        <exclusion>
            <groupId>org.mongodb</groupId>
            <artifactId>mongo-java-driver</artifactId>
        </exclusion>
    </exclusions>
</dependency>
<dependency>
    <groupId>com.querydsl</groupId>
    <artifactId>querydsl-apt</artifactId>
    <version>5.0.0</version>
</dependency>
```

Apart from these dependencies, we need to add build-plugins in the pom.xml file. These plugins help generate Q classes dynamically, which are required for QueryDSL to work properly. The following plugins need to be added:

```xml
<plugin>
    <groupId>com.mysema.maven</groupId>
    <artifactId>apt-maven-plugin</artifactId>
    <version>1.1.3</version>
    <dependencies>
```

```xml
        <dependency>
            <groupId>com.querydsl</groupId>
            <artifactId>querydsl-apt</artifactId>
            <version>5.0.0</version>
        </dependency>
    </dependencies>
    <executions>
        <execution>
            <phase>generate-sources</phase>
            <goals>
                <goal>process</goal>
            </goals>
            <configuration>
                <outputDirectory>target/generated-sources/apt</
outputDirectory>
                <processor>org.springframework.data.mongodb.
repository.support.MongoAnnotationProcessor</processor>
                <logOnlyOnError>false</logOnlyOnError>
            </configuration>
        </execution>
    </executions>
</plugin>
```

GraphQL-related dependencies also need to be added to the POM file of the project. The following GraphQL dependencies need to be added:

```xml
<!-- GraphQL dependencies -->
<dependency>
    <groupId>com.graphql-java</groupId>
    <artifactId>graphql-java</artifactId>
    <version>11.0</version>
</dependency>

<dependency>
    <groupId>com.graphql-java</groupId>
    <artifactId>graphql-java-spring-boot-starter-webmvc</
artifactId>
```

```
    <version>1.0</version>
</dependency>
```

As shown in the preceding code block, to implement GraphQL server in Java, we need to import the `graphql-java` dependencies and the Spring Boot starter for GraphQL JAR file called `graphql-java-spring-boot-starter-webmvc`.

Now that we have added all the necessary dependencies, we will create the entry point, or the `Main` class, of our Spring Boot application, as follows:

```
@SpringBootApplication
public class GraphqlDaaSApp {

    public static void main(String[] args) {
        SpringApplication.run(GraphqlDaaSApp.class,args);
    }
}
```

First, we will create the **Data Access Object** (**DAO**) layer. The DAO layer is the same as it is for the REST application, as described in *Chapter 9, Exposing MongoDB Data as a Service*. In the DAO layer, we will create the `MongoConfig` class, which creates two Mongo Spring beans of the `MongoClient` and `MongoTemplate` type, as shown in the following code block:

```
@Bean
public MongoClient mongo() throws Exception {
    final ConnectionString connectionString = new
ConnectionString(mongoUrl);
    final MongoClientSettings mongoClientSettings
= MongoClientSettings.builder().
applyConnectionString(connectionString).serverApi(ServerApi.
builder()
            .version(ServerApiVersion.V1)
            .build()).build();
    return MongoClients.create(mongoClientSettings);
}


@Bean
public MongoTemplate mongoTemplate() throws Exception {
```

```
        return new MongoTemplate(mongo(), mongoDb);
}
```

Now, we will create a POJO class called `Application`, which represents the data model. It should be annotated by `org.springframework.data.mongodb.core.mapping.Document` annotation and `com.querydsl.core.annotations.QueryEntity` annotation. The following code denotes the `Application` bean:

```
@QueryEntity
@Document(collection = "newloanrequest")
public class Application {
    @Id
    private String _id;
    private String applicationId;
```

Here, `@Document` denotes that the POJO is a bean that is mapped to a document in MongoDB, while `@QueryEntity` is required to enable QueryDSL's querying capabilities over the `Application` bean.

Now, just like the REST-based solution discussed in *Chapter 9, Exposing MongoDB Data as a Service*, we must create an `ApplicationRepository` interface that extends the `MongoRepository` and `QuerydslPredicateExecutor` interfaces. Using this class, we will define two methods to fetch application data from MongoDB using QueryDSL. The following code snippet is for the `ApplicationRepository` class:

```
public interface ApplicationRepository
extends MongoRepository<Application, String>,
QuerydslPredicateExecutor<Application> {

    @Query(value = "{ 'applicationId' : ?0 }")
    Application findApplicationsById(String applicationId);
    @Query(value = "{ 'id' : ?0 }")
    List<Application> findApplicationsByCustomerId(String id);
}
```

We will skip explaining this repository interface as it is identical to the interface we created in the previous chapter.

Now that we have finished developing the DAO layer, let's create a package called `helper`. In a GraphQL Java application, we need two kinds of classes – one should be the GraphQL provider and the other should be the GraphQL data fetcher. Here, we will start by writing the provider class under the `helper` package. In the `GraphQLProvider` class, first, we will define a property of the `graphql.GraphQL` type and initialize it as soon as the `GraphQLProvider` bean is initialized by Spring Boot. The code snippet for this is as follows:

```
...
import graphql.GraphQL;
...
@Component
public class GraphQLProvider {
...
private GraphQL graphQL;
@PostConstruct
public void init() throws IOException {
    URL url = Resources.getResource("schema.graphqls");
    String sdl = Resources.toString(url, Charsets.UTF_8);
    GraphQLSchema graphQLSchema = buildSchema(sdl);
    this.graphQL = GraphQL.newGraphQL(graphQLSchema).build();
}
private GraphQLSchema buildSchema(String sdl) {
    TypeDefinitionRegistry typeRegistry = new SchemaParser().
parse(sdl);
    RuntimeWiring runtimeWiring = buildWiring();
    SchemaGenerator schemaGenerator = new SchemaGenerator();
    return schemaGenerator.makeExecutableSchema(typeRegistry,
runtimeWiring);
}


private RuntimeWiring buildWiring() {
    return RuntimeWiring.newRuntimeWiring()
            .type(newTypeWiring("Query")
                    .dataFetcher("byApplicationId",
graphQLDataFetchers.getApplicationbyApplicationIdDataFetcher())
                    .
dataFetcher("byCustomerId",graphQLDataFetchers.
getApplicationsbyCustomerIdDataFetcher()))
            .build();
}
```

A GraphQL server should have a strongly typed well-defined schema (please refer to the earlier discussion about the GraphQL schema). Here, in the `init` method, we load the GraphQL schema from the resources. The GraphQL schema definition is read and stored in the string `object named sdl` using the utility methods of `com.google.common.io.Resources`. Then, the `GraphQLSchema` object is derived from the `sdl` object, which is built using the `buildSchema` method.

In the `buildSchema` method, the `sdl` object is parsed using `SchemaParser` and converted into the `TypeDefinitionRegistry` object. Runtime wiring is the process of attaching data fetchers, type resolvers, and custom scalars. First, we build the wiring required to complete the GraphQL schema using the `buildWiring` method. Then, using `SchemaGenerator.makeExecutableSchema`, we create a `GraphQLSchema` object with the required `runtimeWiring`.

> **Necessary reference**
>
> Usually, to completely create an executable GraphQL schema, three kinds of bindings might be required. They are as follows:
>
> - **Data fetchers**, which are interfaces that help fetch data for a GraphQL schema.
>
> - Type resolvers, which are custom methods that resolve the value of a GraphQL field.
>
> - Custom scalars, which refer to any customized data types. Data types in GraphQL are known as **scalars**.

Now, let's discuss the implementation of the `buildWiring` method. This method creates a new `Runtimewiring` by attaching GraphQL data fetchers to the two different fields (`byApplicationId` and `byCustomerId`) in the GraphQL schema (the GraphQL schema for this use case was discussed earlier in this chapter).

Finally, using this `GraphQLSchema`, we build and instantiate the GraphQL property. Now, we can expose a bean called `GraphQL` by returning this GraphQL object, as shown here:

```
@Bean
public GraphQL graphQL() {
    return graphQL;
}
```

While implementing the `buildWiring` method, two methods from the `GraphQLDataFetcher` class called `getApplicationbyApplicationIdDataFetcher` and `getApplicationsbyCustomerIdDataFetcher` are called. So, let's discuss how the `GraphQLDatafetcher` class is implemented. All the data fetcher methods must return an object of the `graphql.schema.DataFetcher` type. The definition of the `DataFetcher` interface looks as follows:

```
@PublicSpi
public interface DataFetcher<T> {
    T get(DataFetchingEnvironment var1) throws Exception;
}
```

As shown in the preceding code block, the DataFetcher interface has only one method, which takes an argument of the graphql.schema.DataFetchingEnvironment type. So, we can implement this interface as a Lambda function in Java. In our case, we call the applicationRepository class to fetch the data required for populating the Application object(s) we are publishing. The following code shows the implementation of getApplicationbyApplicationIdDataFetcher:

```
public DataFetcher getApplicationbyApplicationIdDataFetcher() {
    return dataFetchingEnvironment -> {
        String applicationId = dataFetchingEnvironment.
getArgument("applicationId");
        return applicationRepository.
findApplicationsById(applicationId);
    };
}
```

While implementing the getApplicationbyApplicationIdDataFetcher method, we are returning the Lambda function, which takes the dataFetchingEnvironment argument. All the arguments of the field that a data fetcher is written for can be accessed by the getArgument method of the DataFetchingEnvironment class. In this scenario, it is fetching the applicationId argument. Finally, as evident from the code snippet, we use applicationRepository to fetch data from MongoDB. Similar logic is used for writing the getApplicationsbyCustomerIdDataFetcher method.

Now, we need to keep the schema.graphqls file, which contains the GraphQL schema, in the resource folder.

Finally, we need to define the application.yaml file to run the Spring Boot application. The application.yaml file looks as follows:

```
grahqldaas:
  mongoUrl: mongodb+srv://<mongodburl>/
CRRD?retryWrites=true&w=majority
  mongoDb: CRRD
```

Now that we have developed the GraphQL Server application, let's explore how can we test it.

First, run the application by running the `Main` class, as shown in the following screenshot:

Figure 10.11 – Running the GraphQL server application

Now, to test the application, open GraphQL Playground and enter the DaaS endpoint. The DaaS endpoint should look as follows:

```
http://<host_name>:8080/graphql
```

Once you hit this URL on GraphQL Playground and enter a proper `graphql` request payload, you will get the result in GraphQL Playground, as shown in the following screenshot:

Figure 10.12 – Testing the GraphQL DaaS using GraphQL Playground

As we can see, while calling the GraphQL Service application, the client sends the list of fields it wants to fetch. As shown in the request payload, although the GraphQL Server application supports many more fields, in this example, the client has only requested four fields – that is, `applicationId`, `genderCode`, `id`, and `riskScore`. Hence, GraphQL resolves and sends only those four fields back to the consumer. This feature of GraphQL helps overcome the problem of over-fetching or under-fetching (typically seen in REST-based DaaS).

Also, GraphQL supports multiple functionalities in a single call. In our use case, we have two functions – fetch application by `applicationId` and fetch all the applications for a customer using `customerid`. Both can be achieved in a single call using GraphQL. The following screenshot shows an example of this:

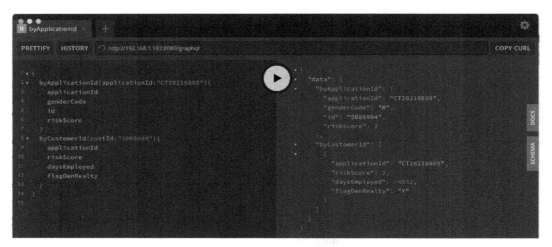

Figure 10.13 – Supporting multiple business operations in a single GraphQL call

As we can see, both fields – `byApplicationId` and `byCustomerId` – can be requested in a single GraphQL call that fetches data for both fields in a single GraphQL response. This reduces the number of hits to the GraphQL service and clients can improve their application performance by making fewer calls to the GraphQL Server application.

Apart from these two advantages, GraphQL also enables easy schema and document sharing. As shown in the preceding screenshot, there are two sidewise tabs called **DOCS** and **SCHEMA** to the extreme right. **SCHEMA** can show us the supported GraphQL schema from the server. It tells us what data points are available as a part of this GraphQL API. The following screenshot shows how a client can see the schema of a GraphQL service:

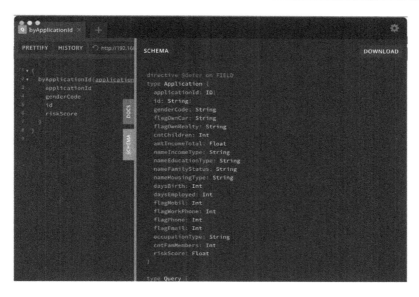

Figure 10.14 – Checking the GraphQL schema from GraphQL Playground

Apart from this, documentation is another cool feature that GraphQL provides. In the world of REST, sharing the API with the client is not sufficient. Therefore, we need to build and maintain the documentation separately (either using Swagger or otherwise) and share it. However, GraphQL allows you to easily maintain and publish documents, as shown in the following screenshot:

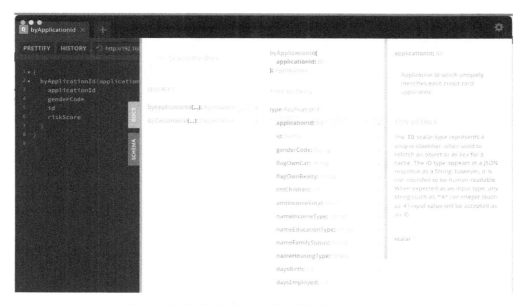

Figure 10.15 – Easily sharing GraphQL documentation

What's more interesting about GraphQL documentation is that it is very easy to configure and maintain. All we need to do is add a documentation comment above each field and query it to support the documentation. For example, the documentation of the `applicationId` field can easily be incorporated into the GraphQL Schema SDL, as shown in the following code block:

```
. . .
"""

Type Application represents the entity/schema of the response
payload for both byApplicationId and byCustomerId fields
"""

type Application {

    """

    Application Id which uniquely identifies each credit card
application

    """

    applicationId: ID!

. . .
```

As we can see, all the documentation is written inside a documentation comment, which starts with three double quotes (" ") and ends with three double quotes (" "). GraphQL automatically uses the schema SDL to publish the documentation, hence making maintenance and publishing documentation super simple and easy.

In this section, we learned how to develop a GraphQL server application using Java. We also learned about GraphQL DSL for developing GraphQL schemas. Finally, we learned how to effectively test and verify the GraphQL DaaS API using GraphQL Playground. Now, let's summarize what we learned in this chapter.

Summary

In this chapter, we learned about the basic concepts of GraphQL. First, we learned how GraphQL can overcome the pitfalls of REST-based DaaS and the benefits it provides. Then, we discussed when to choose GraphQL as a preferred solution. Finally, we learned how to develop a GraphQL server application using Java and how to test that application using GraphQL Playground.

Now that we have learned how to design and develop data engineering solutions for both data ingestion scenarios and data publication scenarios, in the next chapter, we will discuss performance engineering and learn how to use a data-driven approach to make architectural decisions.

Section 4 – Choosing Suitable Data Architecture

In the final section of the book, you will learn how to measure a solution and determine the efficiency of a solution. You will also learn how to communicate and present their solution to leadership/clients who may not be very technical.

This section comprises the following chapters:

11
Measuring Performance and Benchmarking Your Applications

In the preceding chapters, we learned how to architect a solution for data ingestion and data publishing problems. We also discussed how we can choose the correct technology stack and platform to implement a cost-effective and scalable solution. Apart from these, we learned about various architectural patterns for data ingestion. We also discussed data governance and data security. However, as an architect, our job is not only to create a scalable solution but a high-performing one. This is where the role of performance engineering comes into a data architect's toolkit.

In this chapter, we will discuss the meaning of performance engineering and why is it so important. We will also learn how is it different from performance testing. Then, we will learn how to plan our performance tests and other performance engineering activities. Then, we will briefly discuss performance benchmarking techniques. Finally, we will learn about the common methodologies to fine-tune the performance of our solution to mitigate or avoid various kinds of performance bottlenecks during data ingestion or data publishing.

By the end of this chapter, you will know what performance engineering is and how to plan for it. You will know how to benchmark and publish performance results. You will know what the available performance tools are and when to use them. Finally, you will know how to fine-tune performance to create an optimized, highly performant solution for the data problem, as well as how to perform performance benchmarking.

In this chapter, we're going to cover the following main topics:

- Performance engineering and planning
- Tools for performance engineering
- Publishing performance benchmarks
- Optimizing performance

Performance engineering and planning

Software performance engineering (**SPE**) is a systematic and quantitative software-based approach to designing, architecting, and implementing solutions optimally to meet various **non-functional requirements** (**NFRs**) such as performance, capacity, scalability, availability, and reliability. Earlier in this book, we dealt with scalability, availability, and reliability. In this chapter, we will focus on performance and capacity. Alternatively, SPE is defined as a proactive and continuous process of performance testing and monitoring. It involves different stakeholders such as testers, developers, performance engineers, business analysts, and architects. As we will discuss later in this chapter, performance engineering is a seamless process that runs in parallel with development activities, providing a continuous feedback loop to the developers and architects so that performance requirements are imbibed while the software is developed.

Now that we've defined performance engineering, let's discuss the phases of a performance engineering life cycle and how it progresses alongside the **software development life cycle** (**SDLC**) activities. The following diagram shows this:

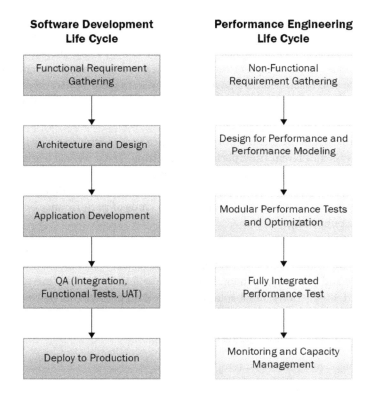

Figure 11.1 – Performance engineering life cycle

As we can see, the following are the various stages of the performance engineering life cycle:

- **NFR gathering**: To develop a high-performant data pipeline, it is important to understand the NFRs of the solution. For example, while designing a DaaS, it makes more sense to know what the required **transactions per second** (**TPS**) are and what the average parallel load to the system would be. There may be another requirement while designing the pipeline, which is that it should be able to be integrated with Datadog for monitoring purposes. In such a scenario, a technology stack should be chosen so that Datadog integration is supported. To gather all this information, we need a close connection between product owners, system analysts, **subject matter experts** (**SMEs**), architects, the SPE team, and DevOps.

- **Design for performance and performance modeling**: In the waterfall model, performance testing and optimization are done after the functional and integrational test cycles are over. The problem with this approach is that sometimes, architectures that work beautifully with small datasets may not work at all in terms of load testing. Due to this, we have to re-engineer the solution again. This causes a lot of waste in terms of effort, time, and money. As modern data engineering teams have increasingly adopted agile methodologies, the opportunity has increased for simultaneously adopting performance engineering. After the nonfunctional requirements have been gathered, designing for performance requires that the following criteria be fulfilled:

  - Satisfy the NFRs with optimal speed

  - The solution must be scalable enough to have similar performance, even with increased load

Our architecture should be designed to scale as business data increases exponentially. This brings to life ideas such as design to fail, scaling out rather than scaling up the application, and auto-scaling resources in the cloud. Another method that's commonly applied during performance-oriented design is **performance modeling**.

Performance modeling is the process of modeling application performance based on features involved in the growth rate of data to find out probable breaches of SLA. It also helps validate design decisions and infrastructure decisions. Let's look at an example – suppose input messages are coming in an application at a rate of x, and the service rate of the application is y. What happens if the arrival rate of messages quadruples itself? How do we ensure the same response time? Do we need to quadruple the service rate or double the service rate? This kind of decision can be made by doing performance modeling. In modern data architectures where applications tend to run on containerized or virtualized platforms, this can determine how to allocate resources to scale in the future.

- **Modular performance tests and optimization**: In this phase, using NFRs, **non-functional tests** (**NFTs**) are classified as stress tests, load tests, or soak tests. Once the test cases have been classified, a test case document that maps NFTs with detailed steps to run the test scenario is prepared. Optionally, an NFT to NFR performance matrix is created. Then, using these documents, NFRs can be tested as a module gets developed and functionally tested. Running these test cases along with functional testing ensures early detection of any performance glitches. Optimization and performance tuning can be done as required by the development or DevOps teams.

- **Fully integrated performance test**: Once the modular performance test is done, end-to-end performance tests can be run. As the **quality assurance** (**QA**) of a scenario finishes, that scenario moves to be tested for integration performance. Usually, in this layer, not much tuning or optimization is required. However, optimization of the overall end-to-end flow may be required during this activity.

- **Monitoring and capacity management**: Once the pipeline is in production, we need to continuously monitor and evaluate any unusual activities. Based on the future and current state of the workloads, capacity can be predicted and managed properly.

In this section, we learned the various stages of the performance engineering life cycle. Next, let's understand the differences between performance engineering and performance testing.

Performance engineering versus performance testing

The key differences between performance engineering and performance testing are as follows:

- Performance testing is a QA activity that runs test cases to check the quality of NFRs and find any issues. It is performed to check how the system will behave in terms of production load and anticipates any issues that could come up during heavy loads. On the other hand, performance engineering is a holistic process that runs hand-in-hand with SDLC. Unlike performance testing, performance engineering starts as early as the analysis phase. It also facilitates the discovery of performance issues early in the development life cycle.

- Performance testing follows the waterfall model of the software development process. It is only done when software development and functional testing have been completed. The problem with such an approach is that if the application fails to perform with production loads, we may need to redesign and re-implement, causing unnecessary time and financial loss. However, performance engineering is a continuous process that goes hand-in-hand with all phases of SDLC and is usually implemented by agile teams with a continuous feedback loop to the development and design team. By providing early analysis of performance needs and early discovery of issues, performance engineering helps us save time and money.

- Performance testing is conducted by the QA team, whereas performance engineering involves architects, developers, SMEs, performance engineers, and QA.

In this section, we learned about performance engineering, why is it needed, and its life cycle. We also discussed the difference between performance testing and performance engineering. In the next section, we will briefly discuss the performance engineering tools available in the market.

Tools for performance engineering

In this section, we will briefly discuss various performance engineering tools.

The following are the different categories of performance engineering tools available:

- **Observability tools**: These tools monitor and gather information about the application. These tools potentially help to identify bottlenecks, track throughput and latency, memory usage, and so on. In data engineering, each system is different, and the throughput and latency requirements are also different. Observability tools help identify if our application is lagging in terms of throughput or latency and by how much. They also help identify hidden issues that may only show up in the long run, in production. For example, a small memory leak in the application may not be noticeable within a few days of deployment. When such an application keeps on running, the tenured region of JVM heap space keeps slowly increasing until it overruns the heap space. The following are a few examples of observability tools:

  - **Datadog**: This is a very popular monitoring tool that can do application monitoring, network monitoring, database monitoring, container monitoring, serverless monitoring, and more. It has an inbuilt dashboard and capabilities to customize your dashboard according to your needs. It has alerting, log integration, and other cool features. It is a paid product, which provides enterprise support. For more information, please visit `https://www.datadoghq.com/`.

  - **Grafana with Graphite/Prometheus**: This is an open source monitoring and dashboarding tool. It is either used with Prometheus or Graphite. Both Prometheus and Graphite are open source monitoring toolkits that help generate and publish various metrics. Prometheus has a data collector module that can pull data to generate metrics. On the other hand, Graphite can only passively listen for data but can't collect it. Some other tool, such as Collectd, needs to collect and push data to Graphite. To query Graphite metrics, functions are used, whereas PromQL is used to query Prometheus metrics. These generated metrics are integrated with Grafana to create different kinds of dashboards, such as stats dashboards, time series monitoring, status timeline and history, alerting dashboards, and so on. For more information, please visit `https://grafana.com/`.

  - **Dynatrace**: Dynatrace is another commercial monitoring and dashboarding tool that has very similar features to Datadog. It also provides an AI assistant to help answer your queries dynamically. It supports DevOps and CloudOps integrations such as CI/CD pipelines and so on. For more information, please visit `https://www.dynatrace.com/`.

  - **Confluent Control Center**: This is an inbuilt confluent Kafka monitoring tool that is shipped with an Enterprise (Licensed) version of Confluent Kafka. It helps monitor various Kafka components such as topics, producers, consumers, Kafka Connect clusters, KSQL queries, as well as overall Kafka cluster health. For more information, please visit `https://docs.confluent.io/platform/current/control-center/index.html`.

  - **Lenses**: This is a tool that provides observability of Kafka topics, clusters, and streams. Lenses not only supports observability but also DataOps for Kafka clusters. For more information, please visit `https://docs.lenses.io/`.

- **Performance testing and benchmarking tools**: These tools are used to do all kinds of performance tests, such as smoke tests, load tests, and stress tests. Some of them also provide benchmarking features. The following are a few of the tools that can be used for performance testing and benchmarking:

  - **JMeter**: This is a free open source tool written in Java that does performance tests. It is especially helpful for big data performance testing and any performance testing of APIs, such as REST and GraphQL. JMeter Hadoop plugins are available to do big data performance testing. We can run the load tests and export the result to a file in multiple formats. For more information, please visit `https://jmeter.apache.org/`.

  - **SoapUI**: This is another open source performance test tool that is used for functional testing as well. It supports load testing with multiple users, threads, and parallelism for web services such as REST, SOAP, and GraphQL. It has a professional commercial edition as well called ReadyAPI. The commercial edition supports more advanced features and specific plugins for testing GraphQL and Kafka streaming applications. For more information, please visit `https://www.soapui.org/`.

  - **Blazemeter**: This is another open source performance test tool that's used to run scalable tests for microservices such as REST or GraphQL APIs. It supports a few monitoring functionalities as well. For more information, please visit `https://www.blazemeter.com/`.

  - **LoadRunner**: LoadRunner is a commercial product from Microfocus that enables load testing for various workloads and various kinds of applications. It supports testing of over 50 types of applications such as microservices, HTML, MQTT, Oracle, and so on. For more information, please visit `https://www.microfocus.com/en-us/products/loadrunner-professional/overview`.

  - **SandStorm**: This is a commercial benchmarking and enterprise-level performance testing tool. It has huge support for various applications and tools, from JDBC connections to big data testing. It supports NoSQL databases such as Cassandra, HBase, and MongoDB, as well as other big data components such as Hadoop, Elasticsearch, and Solar. It also provides support for messaging platforms such as Kafka and RabbitMQ. For more information, please visit `http://www.sandstormsolution.com/`.

  - **Kafka perf test**: This is an inbuilt performance test tool that gets shipped along with Apache Kafka. Two scripts get shipped along with Apache Kafka: `kafka-producer-perf-test.sh` and `kafka-consumer-perf-test.sh`. While the former script is used to test producer performance, the latter is used to test consumer performance. To learn more about this feature, go to `https://docs.cloudera.com/runtime/7.2.10/kafka-managing/topics/kafka-manage-cli-perf-test.html`.

- **OpenMessaging Benchmark Framework**: This is a set of tools that allows you to benchmark distributed messaging systems over the cloud easily. It supports multiple message platforms such as Apache Kafka, Apache Pulsar, Apache RocketMQ, and so on. For more information, please visit `https://openmessaging.cloud/docs/benchmarks/`.

In this section, we briefly discussed multiple tools that can be used for performance engineering. Now that we have a fair idea of what performance engineering is, how to do it, and the tools we can use for it, let's look at how we can create performance benchmarks using the knowledge that we have gained.

Publishing performance benchmarks

In this section, we will learn about performance benchmarking and how to develop and publish them. We will start by defining what a performance benchmark is. A benchmark in software performance testing is defined as a point of reference against which the quality measures of a software solution can be assessed. It can be used to do a comparative study of different solutions to the same problem or compare software products.

Benchmarks are like stats or metrics to determine the quality of software. Just like in sports such as soccer, each player's worth or quality is determined by various stats such as their overall number of goals scored, number of goals scored per match, number of goals scored tournament-wise, and so on. These stats help compare different players under different specifications. Similarly, benchmarks in the software world help determine the worth of a software product or solution under specific conditions.

Now, let's practically run some performance tests and create a performance benchmark. We will use the REST API that we developed in *Chapter 9, Exposing MongoDB as a Service*, to do the performance testing. We will use JMeter to test and record the performance of the application. We chose JMeter since it is easy to use and is an open source product based on Java.

Follow these steps to do a performance test and benchmark:

1. **Add a thread group**: First, we must add a thread group. To add a thread group, we need to do the following:

    I. Start the JMeter tool.

    II. From the tree in the left pane, select **Test Plan**.

    III. Right-click on **Test Plan**. Then, click **Add** | **Threads (Users)** | **Thread Group** to add a thread group, as shown in the following screenshot:

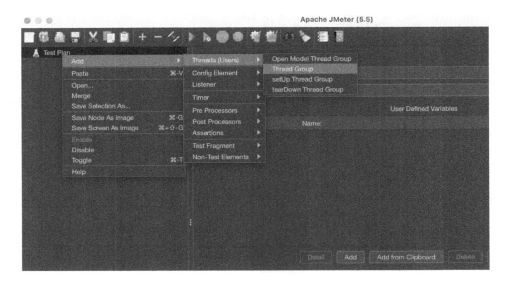

Figure-11.2 – Adding a thread group

In the **Thread Group** creation wizard, fill in the thread group's name and enter the thread's properties, as shown in the following screenshot:

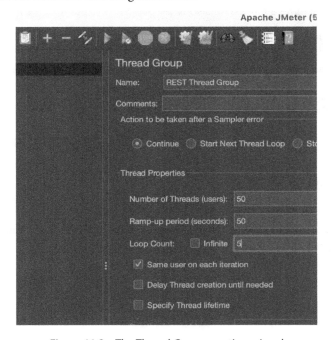

Figure 11.3 – The Thread Group creation wizard

As you can see, **Number of Threads (users)**, **Loop Count**, and **Ramp-up period (seconds)** have been configured. Let's try to understand these terms:

- **Number of Threads (users)** corresponds to the number of users making the request simultaneously, as shown in the following diagram:

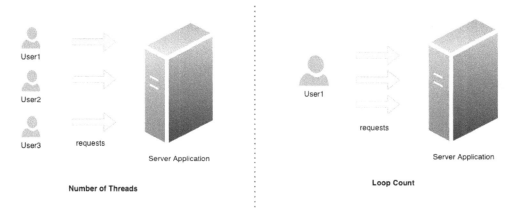

Figure 11.4 – Number of threads versus loop count

- **Loop Count**, on the other hand, corresponds to the number of requests a single user makes to the server. In our example, **Loop Count** is 5. So, a user makes 5 requests to the server.

- **Ramp-up period (seconds)** tells JMeter how much time to wait before starting the next user. In our example, **Number of Threads (users)** is 50 and **Ramp-up period (seconds)** is also 50. Hence, there is a delay of 1 second before it starts the next user.

Now that we have configured our thread group, let's try adding the JMeter elements.

2. **Configure the JMeter elements**: Now, we will add a JMeter config element known as an HTTP Request. This helps make REST or web service calls for load tests. Right-click **REST Thread group** and select **Add | Sampler | HTTP Request**, as shown in the following screenshot:

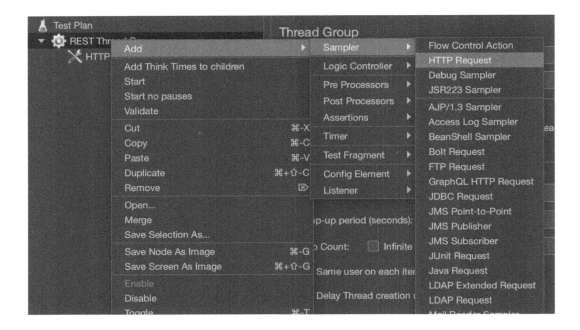

Figure 11.5 – Adding a sampler

Set up **Protocol [http]**, **Server Name or IP**, **Port Number**, **HTTP Request**, and **Path** in the control panel that opens to configure the HTTP Request sampler. We are testing a REST DaaS application, so we will set HTTP **Protocol [http]** to http, **Server Name or IP** to the IP address of the machine where REST DaaS is running, **Port Number** to 8080 (if running on a local machine), **HTTP Request** to GET, and our **Path**, as shown in the following screenshot:

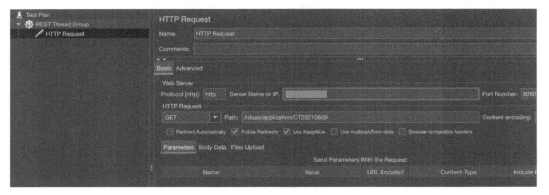

Figure 11.6 – Configuring the HTTP Request sampler

Now that we have configured the HTTP Request sampler, we need to add a few JMeter elements that can monitor and publish performance reports. For this, we must configure one or more listeners.

3. **Add a listener**: To create performance benchmarks, we must add three listeners, as follows:

I.   **Summary Report**

II. **Aggregate Report**

III. **Response Time Graph**

The steps to add a listener are similar to configuring any new listener. Here, we will demonstrate how to add an aggregated performance benchmark, as follows:

IV. Right-click **REST Thread Group** and then select **Add | Listener | Aggregate Report**, as shown in the following screenshot:

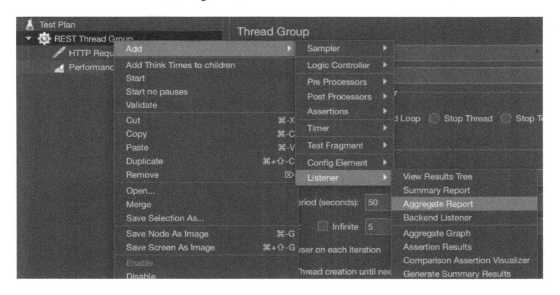

Figure 11.7 – Adding a listener

V.   Rename the report to `Aggregate Performance Benchmark` in the configuration wizard of **Aggregate Report**, as shown in the following screenshot:

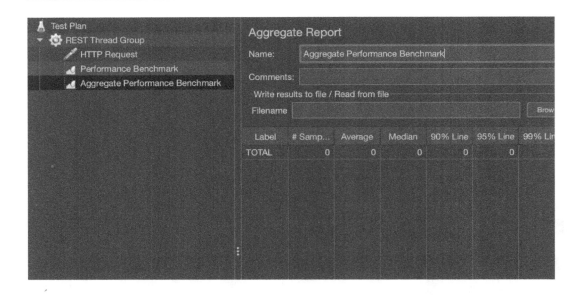

Figure 11.8 – Configuring the Aggregate Report listener

Follow similar steps to set up the **Summary Report** and **Response Time Graph** listeners. By doing so, we will be set to run the test and generate the report.

4. **Run load tests and create a benchmark**: Run the tests by clicking the start symbol that's encircled in red in the following screenshot:

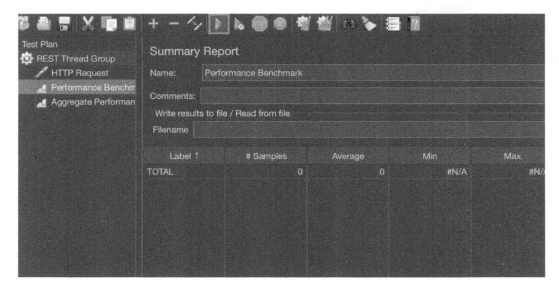

Figure 11.9 – Running performance tests by clicking the Run button

On successfully executing the performance tests, the following reports are generated:

- **Summary Report**: This report provides a summary of the performance benchmark and shows the average, minimum, and maximum response time of the request. You can also see the average throughput of the application. The following screenshot shows a summary of the benchmark results:

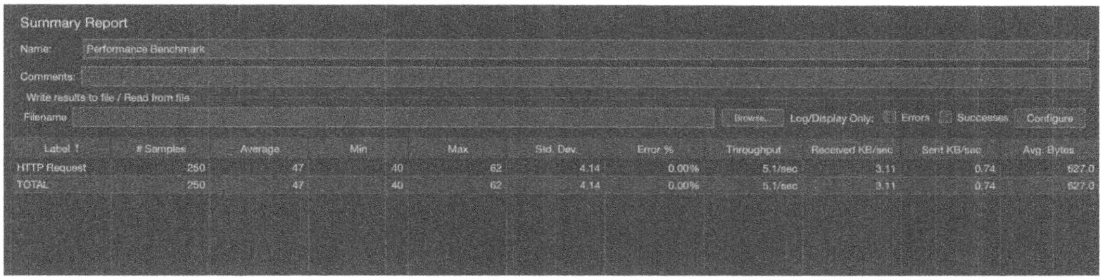

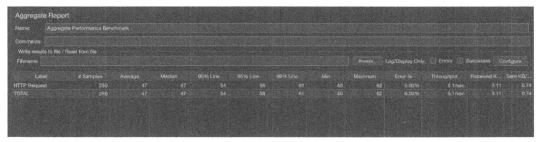

Figure 11.10 – Generated summary benchmark report

Notice the **# Samples** column in the preceding screenshot; its value is **250**. The value of the samples is derived using the following formula:

$$n_{samples} = n_{thread} \times lc$$

$$where\ n_{samples} = Total\ number\ of\ Samples$$

$$n_{thread} = Number\ of\ Threads$$

$$lc = Loop\ count$$

- **Aggregate Report**: This denotes the aggregated benchmark report. Apart from showing the average, median, and maximum response time, it has columns such as **90% Line** and **95% Line**. A **90% Line** column denotes the average response time of 90% of the request. It assumes that 10% of the request contains outliers. Similarly, a **95% Line** assumes that 5% of requests are outliers. The following screenshot shows the aggregated performance benchmark:

Figure 11.11 – Generated aggregated benchmark report

- **Response Time Graph**: A performance benchmark can contain multiple charts or tables to benchmark the application performance. A Response Time Graph depicts the recorded response time at different timelines:

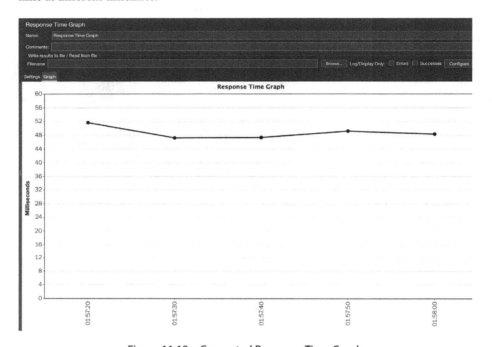

Figure 11.12 – Generated Response Time Graph

In performance benchmarking activities, a lot of the time, we do comparative studies where these reports are used to create a combined report or graph visualization comparing the performance benchmarks of two different solutions.

In this section, we learned why benchmarking is needed and what needs to be considered while benchmarking our solutions. Benchmarking provides us with a way to categorize or classify our application performance as good, bad, or just fine. In the next section, we will find out how can we improve our performance and optimize our data engineering solutions.

Optimizing performance

One of the main reasons that benchmarking, performance testing, and monitoring of applications and systems are done is because of a goal – to optimize the performance so that the system can work to its best potential. The difference between extraordinary software and ordinary software is determined by how well the system is tuned for better performance. In this section, we will learn about various techniques you can use to fine-tune your data engineering pipeline. Although performance tuning is a vast topic, when it comes to various data engineering solutions, we will try to cover the basics of optimizing Java-based data engineering solutions. In the following subsection, we will briefly look at various performance tuning techniques.

Java Virtual Machine and garbage collection optimizations

Java Virtual Machine (**JVM**) performance tuning is the process of adjusting the various JVM arguments or parameters to suit the need of our application so that it performs the best it can.

JVM tuning involves two kinds of optimization, as follows:

- Heap space optimization

- **Garbage collection** (**GC**) optimization

But before we talk about these optimizations, it is worth noting that JVM tuning needs to be the last resort to tune the performance of an application. We should start by tuning application code bases, databases, and resource availability.

Overview of the JVM heap space

Before we deep dive into JVM and GC tuning, let's spend some time understanding the JVM heap space.

The following diagram shows what a JVM heap space looks like:

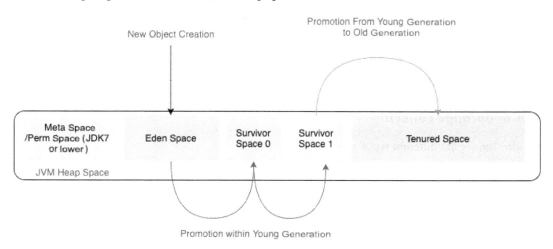

Figure 11.13 – JVM heap space

As we can see, the JVM heap space is divided into four compartments, as follows:

1. Meta or Perm Space

2. Eden Space

3. Survivor Space

4. Tenured Space

Meta Space (known as Perm Space for older JDK versions) stores the metadata of the heap.

Java objects get promoted from Eden Space to Tenured Space, based on the tenure for which they are alive. The following steps show how the Java objects get promoted:

1. The newly created objects from the Java application get stored in *Eden Space*.

2. When *Eden Space* is full, a minor GC event occurs and the objects that are still referenced by the Java application are promoted to *Survivor Space 0*.

3. Again, in the next cycle, when *Eden Space* is full, a second minor GC event gets triggered. At first, this moves all the objects that are still referenced by the application from *Survivor Space 0* to *Survivor Space 1*, and then it promotes referenced objects from *Eden Space* to *Survivor Space 0*.

4. A major GC event occurs when referenced objects are promoted from *Survivor Space 1* to *Tenured Space*. Objects that get promoted to tenured space are called old-generation objects. *Eden Space*, *Survivor Space 0*, and *Survivor Space 1* objects are young-generation objects.

Earlier, we discussed how minor and major GC happens to free up the heap space. But is there a single way or multiple ways to do GC? If so, what should we choose and when? We'll explore this in the next section.

> **Important note**
>
> **Garbage collection** is a process that automatically determines what memory in the JVM is no longer being used by a Java application and recycles that memory for other usages.

Types of garbage collector

The following are the different types of GCs:

*   **Serial garbage collector**: A single GC is suitable for single-threaded applications. It freezes all application threads while doing garbage collection and does so using a single thread.

*   **Parallel garbage collector**: A parallel GC also freezes all threads from the application but uses multiple threads to do garbage collection. Hence, the pause interval of the application threads reduces considerably. It is designed to work for multi-processor environments or multi-threaded environments with medium and large data sizes.

- **Concurrent Mark Sweep (CMS) garbage collector**: As evident from the name, the garbage collection job is performed concurrent to the application. So, it doesn't require application threads to pause. Instead, it shares threads with the application thread for concurrent sweep execution. However, it needs to pause application threads shortly for an **initial mark pause**, where it marks the live objects initially. Then, a second pause, called a **remark pause**, suspends application threads and is used to find any Java objects that need to be collected. These Java objects are created during the concurrent tracing phase. The following diagram explains the difference between serial, parallel, and CMS GCs:

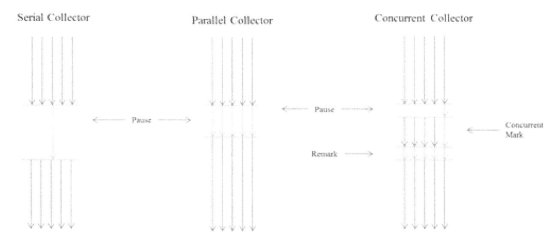

Figure 11.14 – Difference between the single, parallel, and CMS garbage collectors

As we can see, the serial collector uses a single thread for garbage collection, while pausing all application threads. The parallel collector pauses the application threads but since it uses multiple threads to do its job, the pause time is less. CMS, on the other hand, runs concurrently along with the application threads after the initial mark phase.

- **G1 garbage collector**: This is a relatively new GC introduced in Java 7 and later. It depends on a new algorithm for concurrent garbage collection. It runs its longer job alongside the application threads and quicker jobs by pausing the threads. It works using the evacuation style of memory cleaning. For the evacuation style of memory cleaning, the G1 collector divides the heap into regions. Each region is a small, independent heap that can be dynamically assigned to Eden, Survivor, or Tenured Space. The following diagram shows how the G1 collector sees the data:

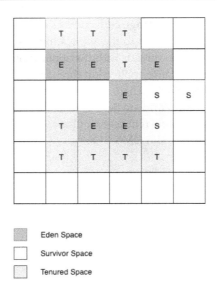

Eden Space
Survivor Space
Tenured Space

Figure 11.15 – G1 garbage collector divides heap space into regions

The GC simply copies data from one region to another. This needs to be retained and marks the older region as blank.

- **Z garbage collector**: This is an experimental GC for very scalable low latency implementations.

The following points should be kept in mind regarding GC:

- Minor GC events should collect as many dead objects as possible to reduce the frequency of full GC.

- More efficient object cleanup is possible when more memory is available for a GC event. More efficient object cleanup ensures a lower frequency of full GC events.

- In the case of performance tuning using GC, you can only tune two parameters out of the three – that is, throughput, latency, and memory usage.

Next, let's see how to tune the performance using a GC.

Tuning performance using a GC

Performance tuning using GC settings is also known as GC tuning. We must follow these steps to perform GC tuning:

1. **Investigate the memory footprint**: One of the most commonly used methods to look for any performance issue caused due to GC can be found in the memory footprint and is present in the GC logs. GC logs can be enabled and generated from a Java application without affecting performance. So, it is a popular tool to investigate performance issues in production. You can enable GC logs using the following command:

```
-XX:+PrintGC -XX:+PrintGCDetails -XX:+PrintGCTimeStamps
-Xloggc:<filename>
```

The following screenshot shows what a typical GC log looks like:

```
2022-07-01T16:45:0.193: [GC (Allocation Failure) [PSYoungGen: 71912K-
>10752K(141824K)] 101680K->101012K(316928K), 0.3575121 secs] [Times:
user=0.22 sys=0.06, real=0.38 secs]
2022-07-01T16:46:0.357: [GC (Allocation Failure) [PSYoungGen: 141832K-
>10752K(141832K)] 232084K->224396K(359424K), 0.569666 secs] [Times:
user=0.45 sys=0.02, real=0.56 secs]
2022-07-01T16:46:0.434: [Full GC (System.gc()) [PSYoungGen: 10752K-
>0K(141824K)] [ParOldGen: 213644K->215361K(459264K)] 224396K-
>215361K(601088K), [Metaspace: 2649K->2649K(1056768K)], 3.4609247 secs]
[Times: user=3.40 sys=0.02, real=3.46 secs]
2022-07-01T16:49:0.984: [GC (Allocation Failure) [PSYoungGen: 131072K-
>10752K(190464K)] 346433K->321225K(649728K), 0.1407158 secs] [Times:
user=0.07 svs=0.08, real=0.14 secs]
```

Figure – 11.16 – Sample GC log

In the preceding screenshot, each line shows various GC information. Let's focus on the third line of the log, which shows a full GC event:

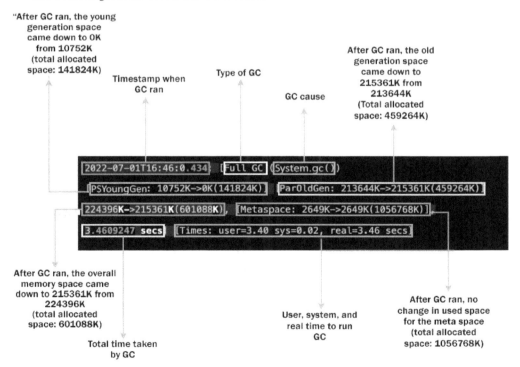

Figure – 11.17 – Anatomy of a GC log statement

As shown in the preceding diagram, let's pick apart the GC log statement and understand the various parts of it:

- `2022-07-01T16:46:0.434`: The timestamp when the GC event occurred.

- `Full GC`: This field describes the type of GC. It can either be a full GC or GC.

- `[PSYoungGen: 10752K->0K(141824K)]`: After the GC event occurred, all the space used in the young generation was recycled. Also, the value inside brackets (`141824K`) denotes the total allocated space in the young generation.

- `[ParOldGen: 213644K->215361K(459264K)]`: After the GC ran, the old generation's used space increased from `213644K` to `215361K`. Total allocated memory for the old generation is `459264K`.

- `224396K->215361K(601088K)`: After the GC ran, the total memory of used space was reduced from `224396K` to `215361K`. The total allocated memory is `601088K`.

- `[Metaspace: 2649K->2649K(1056768K)]`: No memory was reclaimed from the Meta space as a result of the GC. The total allocated Meta space is `1056768K`.

- `3.4609247 secs`: Total time taken by GC.

- `[Times: user=3.40 sys=0.02, real=3.46 secs]`: This part of the log statement tells us about the time taken to do the garbage collection. The `user` time tells the processor time that the GC took to execute. The `sys` time denotes the time taken by I/O and other system activities. Finally, the `real` time denotes the total time taken to finish the GC event.

Based on these footprints, we can determine whether we need to increase the heap space or increase the meta space and specifies any memory leaks that are happening in the application.

2. **Memory tuning**: A memory leak may occur if the following observations are listed:

- The JVM heap size is being filled frequently

- The young generation space is being completely recycled, but the old generation used space is increasing with every GC run

Before deciding whether it is a genuine memory leak problem or not, you should increase the heap space using the following commands:

```
-Xms<heap size>[unit] // for min heap size
-Xmx<heap size>[unit] //for max heap size
//unit can be g(GB),m(MB) or k(KB)
```

If this doesn't help, then the root cause is most likely a memory leak.

If we see that the young generation space is getting filled frequently or that the Meta space is being heavily used, we can plan to change the total allocated space in all these regions using the following commands:

- `-XX:MaxMetaspaceSize`: This sets the maximum amount of memory that can be allocated for the class metadata. The default value is `infinite` (or the same as the heap space).

- `-XX:MetaspaceSize`: Sets the threshold size of the allocated class metadata above which the GC will be triggered.

- `-XX:MinMetaspaceFreeRatio`: The minimum percentage of the Meta space memory region that needs to be available after garbage collection. If the amount of memory left is below the threshold, the Meta space region will be resized.

- `-XX:MaxMetaspaceFreeRatio`: The maximum percentage of the Meta space memory region that needs to be available after garbage collection. If the amount of memory left is above the threshold, the Meta space region will be resized.

- `-XX:NewSize`: This sets the initial size of the young generation space.

- `-XXMaxNewSize`: This specifies the maximum size of young generation space.

- `-Xmn`: This specifies the size of the entire young generation space, meaning Eden and the two survivor spaces.

3. **Latency tuning**: If you come across a scenario where you are seeing that the application is being frozen a lot although CPU utilization or memory utilization is not being used much by the application, then you must tune the GC for latency. Follow these steps to perform latency tuning:

   I. Check the type of GC that the application is using. Change it to G1, if it's not already set. You can enable G1 using the `-XX:+G1GC` command.

   II. Set `-Xmx` and `-Xms` to the same value to reduce application pause intervals.

   III. Set the `-XX:+AlwaysPreTouch` flag to `true` so that the memory pages are loaded when the application is started.

   IV. If you are using G1, check whether the minor GC or full GC is taking more time. If the minor GC is taking more time, we can reduce the values of `-XX:G1NewSizePercent` and `-XX:G1MaxNewSizePercent` . If the major GC is taking more time, we can increase the value of the `-XX:G1MixedGCCountTarget` flag, which will help spread the tenured GC into multiple runs and reduce the frequency of full GC events.

4. **Throughput tuning**: To increase the throughput of an application, you can do either of the following or both:

- Increase the value of the `--XX:MaxGCPauseMillis` property to clean more garbage in a single GC run. However, this may affect your latency.

- Load memory pages into memory at the start of the application by setting the `-XX:+AlwaysPreTouch` and `-XX:+UseLargePages` flags.

> **Important note**
> **Latency** is the total time elapsed to process and send an event, message, or data to its destination. On the other hand, **throughput** is the number of records, events, or messages processed within a specified period.

Although JVM and GC tuning is a vast topic, we briefly tried to cover a few important JVM and GC tuning techniques to improve throughput and latency. In the next section, we will discuss how can we optimize big data loads.

Big data performance tuning

Performance tuning for big data is a huge topic. For brevity, we will limit ourselves to a few performance tuning tips and tricks that are usually applied to the most popular big data processing technologies, namely Spark and Hive.

Hive performance tuning

Hive performance tuning is the collective process and technique to improve and accelerate the performance of your Hive environment. The following are some commonly faced Hive performance issues:

- **Slow-running queries**: Often, you will notice that your Hive query is taking a huge amount of time to finish. There can be several reasons for slow-running queries. The following are a few commonly encountered scenarios of slow-running queries and their solution:

  - Poorly written queries result in a cross-join or full outer join. An unintended cross-join can happen when the join columns in either of the tables have duplicates or a self-join is happening in the query. Fine-tune your Hive queries to avoid cross-joins as much as possible.

  - The speed of a slow-running query can be improved by applying a map-side join if one of the join tables contains a small amount of data. In a map-side join, the smaller dataset is broadcast to all mapper nodes so that the join happens locally without much shuffling. The downside of a map-side join is that the data in the smaller table needs to be small enough to fit into memory.

  - For a scenario suitable for a map-side join, we can further improve the speed of the tables needed for the join to be bucketed. **Bucketing** is the technique by which data in a Hive table is broken into a fixed number of ranges or clusters, based on the join column(s). A bucketed table can be used for a **bucket map-join** or **sort-merge-bucket (SMB) map-join**, both of which perform better than normal map-joins. However, bucketed tables can be joined with each other, only if the total buckets of one table are multiples of the number of buckets in the other table. For example, Table1 has 2 buckets and Table2 has 4 buckets. Since 4 is a multiple of 2, these tables can be joined.

  - Sometimes, data grows rapidly, which makes Hive jobs slow. In such cases, map-side operations take a huge amount of time. To overcome such issues, use partitioning.

- If you notice that data read is slow or data shuffle is quite slow, check for the *data format* and *compression* that were used. For example, columnar structures such as Parquet and ORC have 5% to 10% higher performance than JSON. The columnar format also is known to have around a 90% higher compression ratio than JSON. More compressed data can reduce a good amount of network latency and hence improve overall performance.

- The Hive execution speed can be considerably improved by changing the *execution engine* from map-reduce to Tez or Spark. You can do this by setting the following parameter in the Hive configuration file:

```
<property>
  <name>hive.execution.engine</name>
  <value>spark</value>

</property>
```

- **Job failures**: Job failures are comparatively rare in Hive. Most job failures don't occur due to performance bottlenecks. However, if you observe that the job is running for a long time and hanging and then it fails due to a `TimeoutException`, chances are that you have encountered a *small file issue*. This occurs when millions of small files (whose size is less than the block size of 128 MB) are written into the Hive table's external path. Each HDFS file has metadata stored in Hadoop's name-node. Too many small files in a Hive table cause too much metadata to be read by the job. Hence, the job fails either due to memory overrun or timeout exceptions. In such cases, either run a compaction job (refer to the *Core batch processing patterns* section of *Chapter 7, Core Architectural Design Patterns*) or store the data in sequential files.

With that, we've discussed various Hive optimization techniques that are used to fine-tune Hive query speeds and performance. Now, let's look at Spark performance tuning.

Spark performance tuning

Spark is the most popular big data processing engine. When Spark applications are tuned properly, it lowers the resource cost while maintaining the SLA for critical processes. This is important for both cloud and on-premise environments.

A Spark tuning job starts by debugging and observing the problems that occur during a Spark job's execution. You can observe various metrics using the Spark UI or any profiling application such as Datadog.

The following are a few best practices for Spark optimization that are commonly applied while handling Spark applications:

- **Serialization**:

    - *Problem*: Slow data read from a Hive table or HDFS

- *Cause*: The default Java serializer slows down the read speed of the data from Hive or HDFS

- *Solution*: To overcome this problem, we should set the default serializer to Kyro Serializer, which performs a much faster Serde operation

- **Partition sizes**:

  - *Problem*: The Spark job runs slowly since one or two executors among many executors take much more time to finish the task. Sometimes, the job hangs due to a slow-performing executor.

  - *Cause*: A possible cause can be data skew. In such a scenario, you will find that the slow-performing executor is processing the bulk of the data. Hence, the data is not well-partitioned.

  - *Solution*: Repartition the data loaded into Spark DataFrames before processing the data further.

  Let's look at another common problem that's encountered related to partitioning:

  - *Problem*: The Spark jobs are slowing down as all executors are heavily loaded and all executors are taking a long time to process the job.

  - *Cause*: The likely cause is that the data has outgrown and you are running the Spark job with far fewer partitions.

  - *Solution*: Repartition the data loaded in the Spark DataFrame and increase the number of partitions. Adjust the number of partitions until optimum performance is achieved.

  Another common problem that's encountered in Spark is as follows:

  - *Problem*: A Spark job needs to write a single file as an output but the job gets stuck in the last step.

  - *Cause*: Since you have used `repartition()` to create a single output file, a full data shuffle takes place, choking the performance.

  - *Solution*: Use `coalesce()` instead of `repartition()`. `coalesce()` avoids doing a full shuffle as it collects data from all other partitions and copies data to a selected partition that contains the maximum amount of data.

- **Executor and driver sizing**: Now, let's look at a few performance issues that you can encounter due to ineffective settings for the executor and driver resources:

  - *Problem*: We are encountering multiple memory issues such as `OutofMemoryException`, GC overhead memory exceeded, or JVM heap space overrun.

  - *Cause*: Inefficient configuration of the driver and executor memory and CPU cores.

  - *Solution*: There is no one solution. Here, we will discuss various best practices that can be used to avoid inefficient driver and executor resource configuration. Before we begin, we will assume that you have basic familiarity with Apache Spark and know its basic concepts. If you are new to Apache Spark, you can read this concise blog about the basics of the Spark architecture: `https://www.edureka.co/blog/spark-architecture/`.

- For both executor and driver sizing, there is a total of five properties that we need to set for Spark job optimization. They are as follows:

  - **Executor cores or executor-cores**: According to the official documentation, an optimal executor core ranges from 2 to 5. For an optimized solution with maximum parallelism, we can set this value to 5 by using the following property:

  ```
  --executor-cores 5
  ```

  - **The number of executors or num-executors**: Setting the number of executors is often tricky. If we are using a shared cluster, then we have to note the number of total minimum cores/vCPUs available. If we are using a standalone cluster, then we must note the core size of each node. Let's assume that this number is 16 per node. Here, the total vCPUs will be 48, if the number of nodes is 3. In such a scenario, the aim should be to utilize all 16 vCPUs for the driver and executor. As discussed previously, maximum parallelism can be achieved by setting the executor per core to 5. So, in a node, the maximum number of executors will be 3 (16 cores / 5 executor-cores per executor). So, for three nodes, we will have a total of 9 executors (3 nodes x 3 executors/nodes). However, according to the best practice documentation of Spark, the executor cores per executor should be the same as the driver core. Hence, we would require 1 worker to work out of the 9 workers as the application driver. Hence, the number of executors will become 8 (9 workers – 1 driver worker). As a general formula, you can use the following:

$$Num\ of\ executors = FLOOR(Tn \times Nc/Ec) - 1$$

Here, Nc denotes the number of cores per node, Ec denotes the executor cores per executor, and Tn denotes the total number of nodes.

In this use case, *Nc* is 16, *Ec* is 5, and *Tn* is 3. Hence, we reach the same result of 8 (FLOOR(16*3/5) -1).

With that, we have discussed how to optimally set the driver and executor properties. However, for Spark 3.0 and above running on a YARN cluster, it makes more sense to enable dynamic allocation. Although you can set a minimum and a maximum number of executor cores, you must let the environment itself determine the number of executors needed. Optionally, you may want to set a cap on the maximum number of executors (by using the spark.dynamicAllocation.maxExecutors property) after discussing this with your Hadoop admin.

- **Executor memory or executor-memory**: To calculate the optimal executor memory, we need to understand the total amount of memory that's used by an executor. The total memory that's used by an executor is the total of the executor memory and the memory overhead:

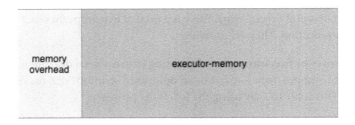

Figure 11.18 – Total memory used by executor = memory overhead + executor-memory

The memory overhead of the executor defaults to a greater amount between 10% of the executor's memory size or 384 MB. Now, to find the correct executor memory size, we need to look at the YARN resource manager's **Nodes** tab. Each record in the **Nodes** tab will have a **Total Memory Column**, as shown in the following screenshot:

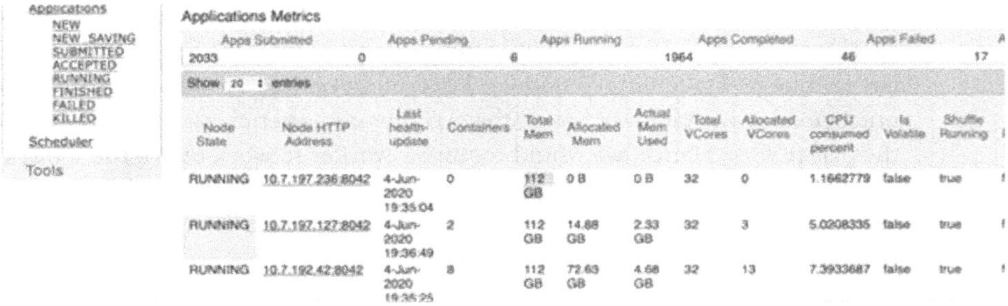

Figure 11.19 – Total memory available for the executors in a node

In the preceding screenshot, a total of **112 GB** of memory is available from the node for executors and drivers, after memory has been set aside for the cluster manager. Now, we must calculate the memory overhead for this executor by using the following formula:

$$memory\ overhead = max\ (384\ MB/1.1 \times M_e\ )$$

$$Where\ M_e = executor\ memory\ per\ node$$

Let's try to use the example described earlier for calculating the executor memory. We will divide this available node memory by the total number of executors per node. Then, we will divide this by 1.1. The formula to calculate executor memory is as follows:

$$executor\ memory = min\ (\frac{Nm}{1.1Ne}, \frac{Nm}{Ne} - .384)\ GB$$

Here, *Nm* is the total node memory and *Ne* is the number of executors per node.

Using the preceding formula in our scenario, the executor memory should be 36.94 GB (112/3 − .384). So, we can set the executor memory to 36 GB.

- **Driver cores or driver-cores**: By default, this will be set to `1`.

- **Driver memory or driver-memory**: Driver memory is either less than or equal to the executor memory. Based on a specific scenario, this value can be set to less than or equal to the executor's memory. One of the optimizations that is advisable in the case of driver memory issues is to make the driver memory the same as the executor memory. This can speed up performance.

- **Directed acyclic graph (DAG) optimization**: Let's look at some issues that you can resolve using DAG optimization:

  - *Problem*: In the DAG, we can see that more than one stage is identical in terms of all its tasks or operations.

  - *Cause*: The DataFrame, denoted as d1, is being used to derive more than one DataFrame (for example – d2, d3, and d4). Since Spark is lazily computed, when creating each DataFrame (d2, d3, and d4), d1 is recalculated every time.

  - *Solution*: In such a scenario, we must persist the dependent DataFrame (d1) using the `Dataset<T> persist(StorageLevel newLevel)` method.

Now that we've discussed performance tuning for big data, let's learn how to tune real-time applications.

Optimizing streaming applications

Now, let's learn how to optimize streaming applications. Here, our discussion will mainly focus on Kafka since we discussed this earlier in this book. When it comes to streaming applications, we can tune their latency and throughput. In this section, we will learn how to observe a performance bottleneck in Kafka. Then, we will learn how to optimize producers and consumers. Finally, we will discuss a few tips and tricks that help tune overall Kafka cluster performance.

Observing performance bottlenecks in streaming applications

The first thing about any tuning is monitoring and finding out what the problem is and where it's occurring. For real-time stream processing applications, this is of utmost importance. There are quite a few Kafka observability tools available such as Lenses and **Confluent Control Center** (**C3**). Here, we will see how C3 helps us observe anomalies.

When you navigate to any topic in C3, you will see three tabs – **Producer**, **Consumer**, and **Consumer lag**. The **Consumer lag** tab can tell if a consumer is slow or not. The following screenshot shows that the consumer (group) is lagging by **1,654** records while reading data from the topic:

Figure 11.20 – Consumer lag

The preceding screenshot also shows the number of active consumers in this consumer group. This can indicate whether the consumer is set to utilize the full performance potential of the topic or not. To learn more, click on the **Consumer group ID** property (in the preceding screenshot, it is **analyticsTeam**). You will see the following screen:

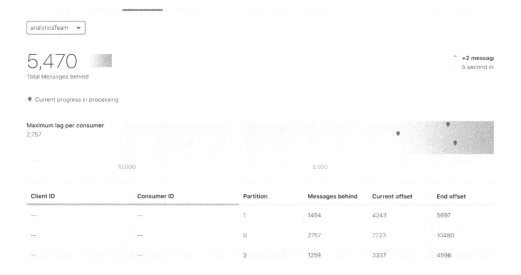

Figure 11.21 – Partition-wise consumer lag

The preceding screenshot shows the lag of the consumer group partition-wise. From *Figure 11.19* and *Figure 11.20*, it is evident that only one consumer is running but that the topic has three partitions. As we can see, there is scope for tuning the consumer application by increasing the number of consumers in the consumer group. Similarly, we can look at the producer speed versus the consumer speed of a topic and find any slowness in the producer.

Producer tuning

The following are a few common tuning techniques for Kafka producers:

- When the producer sends a message to the Kafka broker, it receives an acknowledgment. The `acks=all` property, along with the value of `min.insync.replicas`, determine the throughput of a producer. If `acks=all` is set and the acknowledgment takes more time to come (because it must write all its replicas before sending the acknowledgment), then the producer cannot produce any further messages. This reduces the throughput considerably. We must make a choice here between durability and throughput.

- Idempotent producers guarantee exactly-once delivery of messages. However, this comes with a cost: it reduces the throughput of a producer. So, a system must choose between deduplication and throughput. While using idempotent producers, you can improve throughput slightly by increasing the value of `max.in.flight.requests.per.connection`.

- One way to improve throughput is to increase the property value of `batch.size`. Although a producer is used to send the message, it does so asynchronously (for most applications). Producers usually have a buffer where producing records are batched before they're sent to the Kafka broker. Sending the records to the broker in batches improves the throughput of the producer. However, latency increases as the value of `batch.size` increases. Set `batch.size` to a balanced value so that we get optimum throughput without affecting the latency considerably.

- `Linger.ms` is another property that is affected when the messages in the Kafka producer buffer are sent to the Kafka broker. The higher the value of `linger.ms`, the higher the throughput and latency will be. Again, it must be set in a balanced fashion to have optimum throughput as well as latency. For extremely huge loads of data, a higher value of `linger.ms` can give a considerable boost to performance.

With that, we have briefly discussed the techniques for producer optimization. Now, let's find out how we can optimize the performance of Kafka consumers.

Consumer tuning

The following are a few tips and tricks you can utilize to optimize the consumer to make full use of the potential improvements the consumer can have:

- For a consumer to consume in the most optimized fashion, the total number of consumers should be equal to the number of partitions. In a multithreaded Kafka consumer, the total number of threads across all the consumers in the consumer group should be equal to the number of topic partitions.

- Another typical scenario that makes the consumer slow is too much rebalancing of the consumer. This can happen if the time that's taken to poll and process `max.poll.records` is more than the value of `max.poll.interval.ms`. In such cases, you may want to either increase the value of `max.poll.interval.ms` or decrease the value of `max.poll.records`.

- Rebalancing can happen in scenarios where a consumer can't send the heartbeat or the heartbeat packages reach slowly due to network latency. If we are okay with static consumers (the consumer statically maps to a partition), we can configure a unique `group.instance.id` value for each consumer instance in the consumer group. This will increase latency for the consumer that goes down but will ensure great latency and throughput for other partitions as it will avoid unnecessary consumer rebalancing.

- Since `consumer.commitSync()` blocks your thread unless a commit is done successfully, it may be slower in most cases compared to `consumer.commitAsync()`. However, if there is a fatal error, it makes sense to use `commitSync()` to ensure the messages are committed before the consumer application goes down.

Although we mainly focused this discussion on Apache Kafka, other alternative products that enable stream processing such as Apache Pulsar and AWS Kinesis have similar performance tuning techniques.

Database tuning

Database tuning is an important activity when it comes to performance tuning. It includes SQL tuning, data read and write tuning from database tables, database-level tuning, and making optimizations while creating data models. Since these all vary considerably from one database to another (including SQL and NoSQL databases), database tuning is outside the scope of this book.

Now, let's summarize what we learned in this chapter.

Summary

We started this chapter by understanding what performance engineering is and learning about the performance engineering life cycle. We also pointed out the differences between performance engineering and performance testing. Then, we briefly discussed the various tools that are available to help with performance engineering. We learned about the basics of performance benchmarking and what to consider while creating a benchmark. Then, we learned about various performance optimization techniques and how to apply them to Java applications, big data applications, streaming applications, and databases.

With that, we have learned how to do performance engineering for both batch-based and real-time data engineering problems. In the next chapter, we will learn how to evaluate multiple architectural solutions and how to present the recommendations.

12

Evaluating, Recommending, and Presenting Your Solutions

We started our journey with the basics of data engineering and learned various ways to solve data ingestion and data publishing problems. We learned about various architectural patterns, as well as the governance and security of the solution. In the preceding chapter, we discussed how we can achieve performance engineering and how to create a performance benchmark for our solution.

By now, we have acquired multiple skills to architect efficient, scalable, and optimized data engineering solutions. However, as discussed in the *Responsibilities and challenges of a Java data architect* section of *Chapter 1*, *Basics of Modern Data Architecture*, a data architect has multiple roles to play. In an executive role, the data architect becomes the bridge between business and technology who can communicate ideas effectively and easily with the respective stakeholders. An architect's job is not only to create the solution but also to present and sell their idea to executives and leadership (both functional and technical). In this chapter, we will focus on how to recommend and present a solution.

In this chapter, we will start our discussion by talking about infrastructure and human resource estimation. As data architects, we need to recommend solutions. To do that, we will discuss how we can create a decision matrix to evaluate our solution and compare different alternatives. Then, we will learn about using decision charts effectively to choose an optimal architectural alternative. Finally, we'll learn about a few basic tips and tricks to present the solution effectively.

In this chapter, we're going to cover the following main topics:

- Creating cost and resource estimations
- Creating an architectural decision matrix
- Data-driven architectural decisions to mitigate risk
- Presenting the solution and recommendations

Creating cost and resource estimations

In this section, we will discuss various considerations, methods, and techniques that we apply to create estimations. We will briefly discuss both infrastructure estimations as well as human resource estimations. Infrastructure estimations are closely related to capacity planning. So, we will start our discussion with capacity planning.

Storage and compute capacity planning

To create an infrastructure estimate, you have to find out the data storage needs (RAM, hard disks, volumes, and so on) and the compute needs (the number of CPUs/vCPUs and cores it should have). This process of figuring out the storage and compute needs is called capacity planning. We will start by learning about the factors that need to be considered while conducting capacity planning.

Factors that need to be considered while conducting capacity planning

The following are the various factors that should be taken into consideration while creating a capacity plan:

- **Input data rate**: Based on the type of application, data rates need to be factored in. For example, for a real-time or near-real-time application, the peak data rate should be considered while planning for the storage and compute capacity. For batch-based applications, either the median data rate or the average data rate should be considered. If the batch job runs daily, it is advisable to use median data rates for capacity planning. Median data rates are preferred over average data rates because median data rates are based on the densest distribution of data rates. So, it denotes the middle point of the most frequently recorded data rate. Hence, the median value is never affected by any outlier. On the other hand, the average data rate finds the average of all the data rates over time, including a few exceptional high or low data rates.

- **Data replication and RAID configurations**: Replication ensures high availability and data locality. Since it replicates the data to multiple nodes, systems, or partitions, we must consider the replication factor as well while planning for storage capacity. For example, if 5 GB of data is stored with a replication factor of 3, this means it stores two replicas in different systems, along with the original message. So, the total storage requirement to store 5 GB of data is 15 GB. The replication factor is often mistaken for RAID. While the replication factor ensures high availability using data locality, RAID ensures data safety at the physical storage level by ensuring redundancy at the storage array layer. For mission-critical use cases, it is advisable to take both replication and RAID into consideration while planning for storage capacity.

- **Data retention**: Another important factor is data retention. Data retention is the time for which data needs to be retained in storage before it's purged. This plays an important role as it determines how much storage is needed for accumulation purposes. One of the other things that comes into play in the cloud is the need for archival. Does data need to be archived? If that's true, when should it be archived?

Initially, data may be frequently accessed. Then, it can be infrequently accessed, and the company may want to store the data in an archival zone for long-term audit and reporting requirements. Such a scenario can be handled in the cloud by using specific strategies to save money. For example, we can use S3 intelligent tiering, which sends the data automatically from the S3 standard access to S3 infrequent access layers to the S3 Glacier layers based on the access frequency. This reduces **Operating Expenses (OpEx)** costs as you can make a lot of savings as you move your data from the standard access layer to Glacier.

- **Type of data platform**: It also matters whether you are running your application on-premises or in the public cloud. Capacity planning should consider the maximum required capacity while planning for on-premise deployment. But if you are planning for the cloud, it is advisable to go with a median capacity requirement and configure auto-scaling for peak loads. Since the cloud gives you the option of instant scaling as well as paying for only what you use, it makes sense to spin up the resources required to process the usual data volumes.

- **Data growth**: Another thing you must consider is the growth rate of data. Based on various factors, growth rates may vary. It is important to factor in the growth rate since data engineering pipelines are usually long-term investments.

- **Parallel executions in shared mode**: One of the other factors that we must take into account is shared resources and their effect on concurrent executions. For example, it helps us to correctly estimate the resource requirement of a big data cluster if we know that 10 jobs with an average approximate load of 100 GB may run simultaneously on the cluster. Similarly, to estimate the resource requirements of a Kubernetes cluster, we should be aware of the maximum number of pods that will be running simultaneously. This will help determine the size and number of physical machines and VMs you want to spawn.

With that, we have learned about the various factors that need to be considered while doing storage and compute capacity planning. In the next section, we will look through a few examples of how these factors help in capacity planning.

Applying these considerations to calculate the capacity

In this section, we will discuss a few examples where the aforementioned factors are used judiciously to calculate capacity. The following are a few use cases:

- **Example 1**: Consider that we need to create a capacity plan for a big data cluster in an on-premise data center. Here, the input data rate is R records per second and each record is B bytes. The storage requirement for a day (S_{rd}) can be calculated by multiplying R by B and multiplying the result by 86,400 seconds. However, this calculation doesn't include the replication or RAID factors. We must multiply the replication factor (here, RF) by it, as well as the overload factor (here, OF), due to RAID configurations (the overload factor for RAID 0 is 1, RAID 5 is 1.5, and RAID 10 is 2). Hence, the formula to calculate the storage requirements for a day is as follows:

$$S_{rd} = \frac{R \times B \times 86400 \times RF \times OF}{10^9} \; GB$$

But the actual capacity requirement may be more than this. If there is a requirement to retain data for 7 days, we will get the total required storage by multiplying 7 by S_{rd}. Now, let's see how the growth factor can affect capacity planning. Based on the technology stack, volume of data and data access, and read pattern and frequency, we can set the total memory and compute. Let's say that the calculated storage capacity is S_h, memory is M_h, compute is C_h, and **Input/Output Operations per Second** (**IOPS**) is $IOPS_h$. Also, let's say that the growth rate is g per year. So, the final resource requirement for the next year would be as follows:

$$S_{final} = S_h \times (1 + g)$$

$$M_{final} = M_h \times (1 + g)$$

$$C_{final} = C_h \times (1 + g)$$

$$IOPS_{final} = IOPS_h \times (1 + g)$$

In this example, we saw how our factors help size a Hadoop big data cluster. Hardware **Capital Expenditure** (**CapEx**) represents a significant investment upfront and requires recurring OpEx, hence a balance between the two needs to be attained for better TOC.. In the next example, we'll explore how to size a Kafka cluster for real-time loads.

- **Example 2**: In this example, we are trying to predict the capacity of a Kafka cluster that receives 100 messages per second, where the retention period is 1 week and the average message size is 10 KB. Also, all topics have a replication factor of 3. Here, a Kafka cluster contains two primary clusters – a Kafka cluster and a zookeeper cluster. For a zookeeper cluster in production, a dual-core or higher CPU should be used, along with a memory of 16 to 24 GB. 500 GB to 1 TB disk space should be fine for zookeeper nodes. For the Kafka broker nodes, we should run multi-core servers with 12 nodes and higher. It should also support hyperthreading. The usual normal memory requirement for a Kafka broker is between 24 to 32 GB. Now, let's calculate the storage needs for the broker. The following formula helps calculate the storage needs for each node:

$$S_b = \frac{R \times B \times 86400 \times RF \times DR}{10^9 \times n} GB$$

$$where\ R = incoming\ message\ rate$$

$$B = Bytes\ per\ message$$

$$RF = Replication\ Factor$$

$$DR = Data\ Retention\ in\ days$$

$$n = Number\ of\ Kafka\ brokers$$

By applying this formula to our example, we get the storage needs of each broker as 604 GB.

With these examples, we've seen how we can apply various factors to predict the capacity requirement of a solution. This helps create detailed CapEx and OpEx estimations for the business.

Now that we have learned how infrastructure estimations are calculated, we will discuss how to estimate the human resource-related costs and time for executing a project.

Effort and timeline estimation

Apart from the various responsibilities that an architect has to handle, effort and time estimation is an important responsibility for a data architect. Usually, the architect is responsible for creating a high-level estimate at the start of the project's implementation. Considering most teams follow the agile methodology of software development, detailed effort estimation is done by the agile team during the implementation phase. The following activities need to be done to create a good estimate:

- **Create tasks and dependency charts**: First, to create an estimate, we must analyze the solution and divide it into high-level development and quality assurance tasks. We should also factor in all performance engineering tasks while creating the high-level task list. Then, we must create a dependency task list, which will specify whether a particular task is dependent on another task(s) or not. The following table shows one of the formats for creating a task and dependency list:

Task no.	Task name	Dependency
1	Creating Git user registration and a master repository	
2	Creating local repositories on a PC	Task 1
3	Creating a simple Hello World whose output will be shown in Hindi in Java	
4	Creating R2	
5	Reviewing the code of R2	Task 4
6	Pushing the code of R2	Task 2 and Task 5
7	Creating a data model of R2	
8	Creating a data model of R3-b	
9	Reviewing the data model of R3-b	Task 8
10	Pushing the data model of R3-b	Task 1 and Task 8

Figure 12.1 – Sample task and dependency list

In the preceding table, note that task 2 is dependent on task 1 and, similarly, task 6 is dependent on tasks 2 and 5. Note how we can denote such dependencies by adding a dependency column. This dependency matrix helps us understand the risks and dependencies. It also helps us understand how various tasks can run in parallel. This helps create a roadmap for various feature releases.

- **Classify the tasks based on their complexity**: One of the things that an architect must do is classify the tasks into one of three categories: high complexity, medium complexity, and low complexity. However, for some exceptional use cases, more granular complexity levels can be defined.

- **Classify based on technology**: Another classification that helps with estimation is technology. This is because a complex task for a Spark-based job can be different than a complex task for a DataStage-based job. So, the average effort that needs to be spent not only depends on the complexity but also on the technology.

- **Create the estimated time taken for a task**: To create a time estimate, first, we must create a map consisting of the time taken for a specific combination of technology and complexity if it is a technical task. If it is an analytical task, we must create a mapping for the time taken by a task versus its complexity. For example, a complex task for a Spark job may take 8 man-days to finish, while a complex task for an Informatica job may take 5 man-days to finish. Based on such a mapping, we can estimate the total time taken to finish the project in man-days or man-hours. For some agile projects, this effort can be estimated using a point system. For example, a complex analysis task may take 5 points of effort.

- **Create total effort estimates**: Based on the estimation done in the previous steps, we can calculate the total effort by summing up all individual task efforts required to deliver the solution.

- **Add a buffer to estimates**: As mentioned in the book *The Pragmatic Programmer (Hunt et al., 1999)*, we should remember that *Rather than construction, software is more like gardening – it is more organic than concrete. You constantly monitor the health of the garden and make adjustments (to the soil, the plants, the layout) as needed.* Since developing any application or data pipeline is organic, we must add a buffer to our estimates so that any organic changes may be accommodated in the project schedule.

- **Create a product roadmap and timeline for releases**: Based on the total estimate, dependency, risks, and expected delivery range, we can create a roadmap and timeline. It is important to understand the expected delivery timelines to ensure we can do proper resource loading and deliver the project in the time range that the business is looking for. Having said that, a lot of times, the business has unrealistic expectations of delivery timelines. It is the job of an architect (with the help of the project manager and product owner) to communicate and justify the estimates to the business so that both the technical and business teams can come to a common ground in terms of the delivery timeline.

- **List all risks and dependencies alongside the estimate**: While creating an estimate, it is important to list all the risks, dependencies, and assumptions so that all stakeholders are aware of what is being delivered and the risks involved in delivering the solution.

Now that we have learned how to create and document a well-thought effort estimate, we have to figure out the total delivery or development cost of a solution. To do that, we must perform human resource loading. Human resource loading is a process by which we identify how many developers, testers, and analysts with specific skill sets are required to deliver the project in the agreed-upon time. Finding the right mix of people with specific roles is the key to delivering the solution. Then, we assign a specific per-hour cost based on the role, demographics, and technology.

After that, we factor in the number of hours required by a role or resource and multiply it by the assigned rate to reach the total cost of a human resource for the project. By summing up the total cost of each resource, we can find the total development cost (or the labor cost; any infra or license costs are not included).

With that, we have learned how to create cost and resource estimates to implement a solution. Earlier in this book, we learned how to develop different architectures for various kinds of data engineering problems. We also learned how to run performance tests and benchmarks.

In this section, we learned how to create cost and resource estimates to implement a solution. *Is there a way to stitch all this information together to recommend the best-suited solution?* In real-world scenarios, each data engineering problem can be solved by multiple architectural solutions. *How do we know which is the most suitable solution? Is there a logical way to determine the best-suited solution?* Let's find out.

Creating an architectural decision matrix

Concerning data engineering, an architectural decision matrix is a tool that helps architects evaluate the different architectural approaches with clarity and objectivity. A decision matrix is a grid that outlines the various desirable criteria for making architectural decisions. This tool helps rank different architectural alternatives, based on the score for each criterion. Decision matrices are used by other decision-making processes. For example, decision matrices are used by business analysts to analyze and evaluate their options.

The decision matrix, also known as the Pugh matrix, decision grid, or grid analysis, can be used for many types of decision-making processes. However, it is best suited for scenarios where we have to choose one option among a group of alternatives. Since we must choose one architecture for the recommendation, it makes sense to use a decision matrix to arrive at a conclusion. Now, let's discuss a step-by-step guide to creating a decision matrix for architectural decision-making. The steps to create a decision matrix are as follows:

1. **Brainstorm and finalize the various criteria**: To create a decision matrix that can be used for architectural evaluation, it is important to brainstorm and finalize the various criteria that the decision depends on. It is important to involve leadership and business executives as stakeholders in this brainstorming session. If you are an architect from a services firm who is developing a solution for the client, it is important to involve an architect from the client side as well. This step is very important as various criteria and their priorities help narrow down our final recommendation among a set of architectural alternatives.

2. **Create the matrix table**: Next, we should create the decision matrix table, where each row denotes a specific criterion, while each column denotes a specific architectural alternative. These are selected sets of criteria that help us determine the appropriateness of the architecture for the current use case. The following diagram shows what the table will look like:

Architectural Decision Matrix

	Architecture 1	Architecture 2	Architecture 3	Architecture 4
Criteria 1				
Criteria 2				
Criteria 3	Yet to be developed	Yet to be developed	Yet to be developed	Yet to be developed
Criteria 4				
Criteria 5				

Figure 12.2 – Creating the decision matrix table

3. **Assign rank or scale**: Now, we must assign rank or scale to each of the criteria of the architecture. Rank or scale is a relative measure where the higher the scale, the better it fits the criteria. The following diagram shows how scale is assigned to different architectures based on various criteria:

Architectural Decision Matrix

	Architecture 1		Architecture 2		Architecture 3		Architecture 4	
	Scale		Scale		Scale		Scale	
Criteria 1	3		2		4		4	
Criteria 2	3.5		3.5		3		2.5	
Criteria 3	4.5	Yet to be developed	2.5	Yet to be developed	3	Yet to be developed	5	Yet to be developed
Criteria 4	2		4		4		3.5	
Criteria 5	2.5		3.5		3		4.5	

Figure 12.3 – Assigning scale values for each architecture against each criterion

As we can see, different scales are assigned to each architecture against each criterion on a relative scale of 1 to 5. Here, 5 is the highest match possible for the given criterion, while 1 is the lowest match possible for the given criterion. In this example, *Architecture 1* gets a score of *3* in terms of *Scale*. *Architecture 2* gets a score of *2*, while *Architecture 3* and *Architecture 4* get a score of *4* for *Criteria 1*. Hence, *Architecture 3* and *Architecture 4* are the most suitable as far as *Criteria 1* is concerned.

4. **Assign weights**: Next, we must assign weights to each criterion. This will help set the priority for various criteria. The following diagram shows how weights are assigned to each architecture against each criterion:

Architectural Decision Matrix

	Architecture 1			Architecture 2			Architecture 3			Architecture 4		
	Scale	Weight		Scale	Weight		Scale	Weight		Scale	Weight	
Criteria 1	3	2		2	2		4	2		4	2	
Criteria 2	3.5	3		3.5	3		3	3		2.5	3	
Criteria 3	4.5	2	Yet to be developed	2.5	2	Yet to be developed	3	3	Yet to be developed	5	2	Yet to be developed
Criteria 4	2	4		4	4		4	4		3.5	4	
Criteria 5	2.5	4		3.5	4		3	4		4.5	4	

Figure 12.4 – Assigning weights to each criterion

As we can see, the weight that's assigned to each criterion is independent of the architecture. This attaches a priority to each of the criteria. So, the most desirable criteria get the highest preference. The higher the weight, the higher the priority. In this example, *Criteria 1* and *Criteria 2* get the least priority with a priority score of *2*, while *Criteria 4* and *Criteria 5* get the highest priority with a priority score of *4*.

5. **Calculate the score:** The individual scores for each architecture against a criterion are calculated by multiplying scale values by the weight of the criteria. The total desirability score of the architecture is calculated by summing up all the scores of each criterion. The following diagram shows what such a decision matrix looks like:

Architectural Decision Matrix

	Architecture 1			Architecture 2			Architecture 3			Architecture 4		
	Scale	Weight	Score	Scale	Weight		Scale	Weight	Score	Scale	Weight	Score
Criteria 1	3	2	6	2	2	4	4	2	8	4	2	8
Criteria 2	3.5	3	10.5	3.5	3	10.5	3	3	9	2.5	3	7.5
Criteria 3	4.5	2	9	2.5	2	5	3	3	9	5	2	10
Criteria 4	2	4	8	4	4	16	4	4	16	3.5	4	14
Criteria 5	2.5	4	10	3.5	4	14	3	4	12	4.5	4	18
Total Desirability score			43.5			49.5			54			57.5

Figure 12.5 – Example of a completed decision matrix for architectural decisions

As we can see, *Architecture 4* looks like the most desirable solution as it has the highest total desirability score of *57.5*, while *Architecture 1* is the least desirable with a score of *43.5*.

In this section, we learned about how to create a decision matrix. Now, the question is, *is the total desirability score always enough to recommend an architecture?* In the next section, we'll learn how to further evaluate an architecture by using the techniques we learned earlier.

Data-driven architectural decisions to mitigate risk

A decision matrix helps us evaluate the desirability of an architecture. However, it is not always necessary to opt for the architectural option that has the highest desirability score. Sometimes, each criterion needs to have a minimum threshold score for an architecture to be selected. Such scenarios can be handled by a spider chart.

A spider chart, also known as a radar chart, is often used to display data across multiple dimensions. Each dimension is represented by an axis. Usually, the dimensions are quantitative and normalized to match a particular range. Then, each option is plotted against all the dimensions to create a closed polygon structure, as shown in the following diagram:

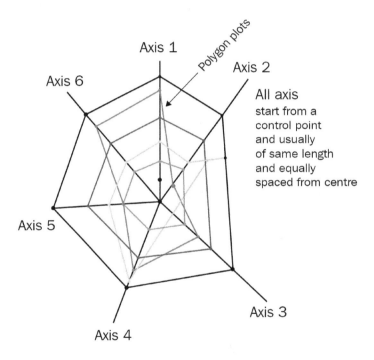

Figure 12.6 – Spider or radar chart

In our case, each criterion for making an architectural decision can be considered a dimension. Also, each architectural alternative is plotted as a graph on the radar chart. Let's look at the use case for the decision matrix shown in *Figure 12.5*. The following diagram shows the radar chart for the same use case:

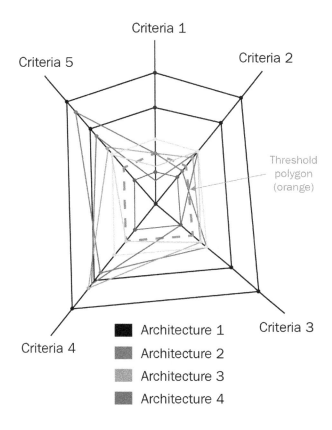

Figure 12.7 – Radar chart for the example scenario discussed earlier

As we can see, each axis denotes a criterion such as *Criteria 1*, *Criteria 2*, and so on. Each criterion has a total score of 25 points, divided into five equal parts along its axis. The division markers for each criterion are connected to the division marker of the adjacent criteria, creating a spider web of symmetrical pentagons. The maximum score of each criterion is 25 because it is the product of the maximum scale value (5) and the maximum weightage (5). We also create a threshold polygon, as denoted by the dotted lines in the preceding diagram. This is created by joining the threshold marker (in this case, a score of 8 points) for every criterion. An optimal solution is one whose polygon is either bigger or equal to the threshold polygon. All the criteria of an optimal solution should score more points than the threshold score of each criterion.

As shown in the preceding diagram, our threshold score for each criterion is 8. Based on the score of each criterion for the architecture, we draw the polygon plot. Here, the plot of *Architecture 1* is blue, *Architecture 2* is pink, *Architecture 3* is green, and *Architecture 4* is violet. Based on the plots, we can see that only *Architecture 3* is optimal in this use case. Although the total desirability score of *Architecture 4* is greater than that of *Architecture 3*, it doesn't fulfill the condition of having the minimum threshold score of 8 for each criterion as it only scores 7.5 in *Criteria 2*. Also, if the individual score of each criterion for *Architecture 3* is more than or equal to the threshold score. Hence, *Architecture 3* is the best-suited option for this use case.

An alternative way to evaluate a decision matrix to find the most optimized solution is using a decision tree. **Decision trees** are decision support tools that use tree-like models for questions and categorize or prune the results based on the answer to those questions. Usually, the leaf nodes denote the category or the decision. The following diagram shows an example of a decision tree for evaluating the scenario that we just discussed using a spider/radar chart:

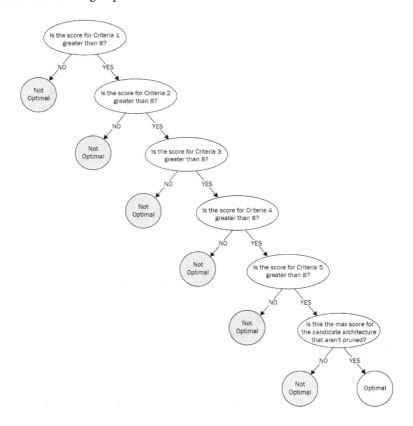

Figure 12.8 – Decision tree to evaluate architectural alternatives

As we can see, we can create a decision tree based on the scores recorded earlier in the decision matrix to find the most optimal solution. Here, we are asking questions such as *Is the score for Criteria 1 greater than 8?*, *Is the score for Criteria 2 greater than 8?*, and so on. Based on the answer, we are pruning out the non-optimal solutions at that level. Finally, we ask, *Is this the max score for the candidate architecture and it can't be pruned?* The answer to this question helps us figure out the most optimal solution.

In this section, we learned how to use a data-driven methodical approach to find and evaluate the most optimal solution for the problem. Now, it is the job of the architect to present the solution as a recommendation. In the next section, we will discuss the guidelines for how to present your solution effectively.

Presenting the solution and recommendations

As an architect, your job doesn't end when you create and evaluate the most optimal architectural alternative based on the problem, platform, criteria, and priorities. As a Janus between business and technology, the architect is also responsible for effectively communicating the solution and recommending the most optimal alternative. Based on the kind of project and the type of business you are in, you may be required to convince stakeholders to invest in the solution. The following are a few guidelines that will help you present your solution and convince your stakeholders more effectively:

- **Present the presentation before the presentation**: If possible, engage the customer or end client early and give them a glimpse of what possible solutions you are thinking of. Also, tell them how much time will it take for you to evaluate each of the solutions. Keep them engaged and in the loop while developing the architecture. It is always helpful if stakeholders are involved in the process and kept in the loop. It's a win-win situation for both the stakeholders and the architect. Stakeholders feel that they are part of the solution, and they get enough time to understand or anticipate any impact of implementing the solution. On the other hand, the architect gets constant feedback about the priorities and criteria, which helps them come up with a very well-researched decision matrix. A more accurate decision matrix eventually helps architects make the most desired recommendation.

- **Know your audience and ensure they are present**: Although this is true for any presentation, it is important to understand the audience before presenting the solution. It is important to understand whether they are from business, executive leadership, or the technical side. It is also important to consider if any external teams or vendors are involved. Understanding the demographics of your audience will help you customize your presentation so that it is relatable to their work. If it is a mixed audience, make sure that you have something relatable for all the different audience groups. It is also important that you invite all the important stakeholders so that your solution reaches every intended audience.

- **Present the Return on Investment (ROI) for the solution**: Usually, there are top-level leaders, executives, and business heads present in a solution presentation. For them, it is important to understand how the solution can either generate or save dollars. It could be that the solution will create additional revenue generation sources, or it could be as trivial as a lesser total cost of ownership or a lesser cost of development. To showcase the ROI for the solution, you can include if the solution adds any value to the customer experience or acceptance of the product. A good data architect should carefully brainstorm and figure out how the solution can add value to the business.

- **Recommend by comparing alternatives**: Although we, as architects, usually recommend one architecture, it is a good practice to present all the alternative architectures and their pros and cons. Then, we must establish the most suitable architecture. It is also a good idea to present why we chose that architecture.

- **Make the presentation better by using their language**: Each company and business has its own language. Since a lot of stakeholders are from the business side of things, it is better to adapt to the common language that's popular in the organization when presenting. This ensures that the audience is easily getting what we are presenting and can connect the dots.

- **Mind the context**: It is also important for the presentation to be contextual. Based on the audience, your presentation should be customized so that it has the correct balance between technical versus business content.

- **Ensure your presentation is visually appealing and relatable**: Diagrams speak more than words. Presentations must have clear diagrams that are relatable and self-explanatory. Avoid too much text in a presentation. A visually appealing presentation is easier to explain and keeps the various stakeholders interested in the presentation.

In this section, we discussed a few tips and tricks for presenting a solution to stakeholders in a concise, effective, and impactful way. Apart from developing and architecting a solution, we are aware of how to evaluate, recommend, and present a solution effectively. Now, let's summarize what we learned in this chapter.

Summary

We started this chapter by learning how to plan and estimate infrastructure resources. Then, we discussed how to do an effort estimation, how to load human resources, and how to calculate the total development cost. By doing so, we learned how to create an architectural decision matrix and how to perform data-driven comparisons between different architectures. Then, we delved into the different ways we can use the decision matrix to evaluate the most optimal solution by using spider/radar charts or decision trees. Finally, we discussed some guidelines and tips for presenting the optimized solution in a more effective and impactful way to various business stakeholders.

Congratulations – you have completed all 12 chapters of this book, where you learned all about a Java data architect's role, the basics of data engineering, how to build solutions for various kinds of data engineering problems, various architectural patterns, data governance and security, and performance engineering and optimization. In this final chapter, you learned how to use data-driven techniques to choose the best-suited architecture and how to present it to the executive leadership. I hope you have learned a lot, and that it will help you develop your career as a successful data architect and help you grow in your current role.

Index

S

Packt.com

Subscribe to our online digital library for full access to over 7,000 books and videos, as well as industry leading tools to help you plan your personal development and advance your career. For more information, please visit our website.

Why subscribe?

- Spend less time learning and more time coding with practical eBooks and Videos from over 4,000 industry professionals

- Improve your learning with Skill Plans built especially for you

- Get a free eBook or video every month

- Fully searchable for easy access to vital information

- Copy and paste, print, and bookmark content

Did you know that Packt offers eBook versions of every book published, with PDF and ePub files available? You can upgrade to the eBook version at packt.com and as a print book customer, you are entitled to a discount on the eBook copy. Get in touch with us at customercare@packtpub.com for more details.

At www.packt.com, you can also read a collection of free technical articles, sign up for a range of free newsletters, and receive exclusive discounts and offers on Packt books and eBooks.

Other Books You May Enjoy

If you enjoyed this book, you may be interested in these other books by Packt:

Data Cleaning and Exploration with Machine Learning

Michael Walker

ISBN: 9781803241678

- Explore essential data cleaning and exploration techniques to be used before running the most popular machine learning algorithms

- Understand how to perform preprocessing and feature selection, and how to set up the data for testing and validation

- Model continuous targets with supervised learning algorithms

- Model binary and multiclass targets with supervised learning algorithms

- Execute clustering and dimension reduction with unsupervised learning algorithms

- Understand how to use regression trees to model a continuous target

Production-Ready Applied Deep Learning

Tomasz Palczewski, Jaejun (Brandon) Lee, Lenin Mookiah

ISBN: 9781803243665

- Understand how to develop a deep learning model using PyTorch and TensorFlow
- Convert a proof-of-concept model into a production-ready application
- Discover how to set up a deep learning pipeline in an efficient way using AWS
- Explore different ways to compress a model for various deployment requirements
- Develop Android and iOS applications that run deep learning on mobile devices
- Monitor a system with a deep learning model in production
- Choose the right system architecture for developing and deploying a model

Packt is searching for authors like you

If you're interested in becoming an author for Packt, please visit `authors.packtpub.com` and apply today. We have worked with thousands of developers and tech professionals, just like you, to help them share their insight with the global tech community. You can make a general application, apply for a specific hot topic that we are recruiting an author for, or submit your own idea.

Share Your Thoughts

Now you've finished *Scalable Data Architecture with Java*, we'd love to hear your thoughts! Scan the QR code below to go straight to the Amazon review page for this book and share your feedback or leave a review on the site that you purchased it from.

`https://packt.link/r/1-801-07308-2`

Your review is important to us and the tech community and will help us make sure we're delivering excellent quality content.